Smart Innovation, Systems and Technologies

Volume 80

Series editors

Robert James Howlett, Bournemouth University and KES International,
Shoreham-by-sea, UK
e-mail: rjhowlett@kesinternational.org

Lakhmi C. Jain, University of Canberra, Canberra, Australia;
Bournemouth University, UK;
KES International, UK
e-mails: jainlc2002@yahoo.co.uk; Lakhmi.Jain@canberra.edu.au

About this Series

The Smart Innovation, Systems and Technologies book series encompasses the topics of knowledge, intelligence, innovation and sustainability. The aim of the series is to make available a platform for the publication of books on all aspects of single and multi-disciplinary research on these themes in order to make the latest results available in a readily-accessible form. Volumes on interdisciplinary research combining two or more of these areas is particularly sought.

The series covers systems and paradigms that employ knowledge and intelligence in a broad sense. Its scope is systems having embedded knowledge and intelligence, which may be applied to the solution of world problems in industry, the environment and the community. It also focusses on the knowledge-transfer methodologies and innovation strategies employed to make this happen effectively. The combination of intelligent systems tools and a broad range of applications introduces a need for a synergy of disciplines from science, technology, business and the humanities. The series will include conference proceedings, edited collections, monographs, handbooks, reference books, and other relevant types of book in areas of science and technology where smart systems and technologies can offer innovative solutions.

High quality content is an essential feature for all book proposals accepted for the series. It is expected that editors of all accepted volumes will ensure that contributions are subjected to an appropriate level of reviewing process and adhere to KES quality principles.

More information about this series at http://www.springer.com/series/8767

Óscar Mealha · Monica Divitini
Matthias Rehm
Editors

Citizen, Territory and Technologies: Smart Learning Contexts and Practices

Proceedings of the 2nd International Conference on Smart Learning Ecosystems and Regional Development - University of Aveiro, Portugal, 22–23, June 2017

 Springer

Editors
Óscar Mealha
Department of Communication and Art
University of Aveiro
Aveiro
Portugal

Monica Divitini
Department of Computer and Information
  Science
Norwegian University of Science
  and Technology
Trondheim
Norway

Matthias Rehm
Department of Architecture, Design,
  and Media Technology
Aalborg University
Aalborg
Denmark

ISSN 2190-3018   ISSN 2190-3026   (electronic)
Smart Innovation, Systems and Technologies
ISBN 978-3-319-87052-6   ISBN 978-3-319-61322-2   (eBook)
DOI 10.1007/978-3-319-61322-2

Printed on acid-free paper

This Springer imprint is published by Springer Nature
The registered company is Springer International Publishing AG
The registered company address is: Gewerbestrasse 11, 6330 Cham, Switzerland

Preface

We are proud to present the proceedings of the second International Conference on Smart Learning Ecosystems and Regional Developments (SLERD2017). Following a successful first edition in Timisoara, Romania, in 2016, the conference was organized in 2017 in Aveiro, Portugal. SLERD2017 was hosted by CIC Digital/DigiMedia Research Group at the University of Aveiro, in the period June 22–23, 2017. DigiMedia[1]—Digital Media and Interaction is an interdisciplinary research group focusing on innovation in the design of new interaction approaches for human-centered digital media applications aiming to foster interpersonal and community-oriented communication.

The conference was co-organized by the ASLERD (Association for Smart Learning Ecosystems and Regional Development[2] an international no-profit inter-disciplinary, democratic, scientific-professional Association that is committed to support learning ecosystems to get smarter and play a central role to regional development and social innovation. "Smart," thus, are not simply technology-enhanced learning ecosystems but, rather, learning ecosystems that promote the multidimensional well-being of all players of learning process (i.e., students, pro-fessors, administrative personnel and technicians, territorial stakeholders, and, for the schools, parents) and that contribute to the increase of the social capital of a "region," also thanks to the mediation of the technologies. ASLERD, thus, aims at generating a concrete impact by understanding learning ecosystems and accom-panying design for "smartness," fostering the development of policies and action plans, supporting technological impact well beyond prototypes and pilots, pro-moting networking and opportunities to discuss and debate like the SLERD yearly conference.

SLERD 2017 aimed at promoting reflection and discussion concerning R&D work, policies, case studies, entrepreneur experiences with a special focus on understanding how relevant the smart learning ecosystems (schools, campus,

---

[1]DigiMedia - http://digimedia.web.ua.pt/.
[2]ASLERD - https://en.wikipedia.org/wiki/ASLERD.

working places, informal learning contexts, etc.) are for regional development and social innovation and how the effectiveness of the relation of citizens and smart ecosystems can be boosted. The conference had a special interest in understanding how technology-mediated instruments can foster the citizen's engagement with learning ecosystems and territories, namely by understanding innovative human-centric design and development models/techniques, education/training practices, informal social learning, innovative citizen-driven policies, technology-mediated experiences and their impact. This set of concerns contributes to foster the social innovation sectors and ICT and economic development and deployment strategies alongside new policies for smarter proactive citizens.

Overall, we received 38 unique submissions from 19 countries, demonstrating the global interest for this research area and for SLERD2017. Out of the total submissions, after a rigorous double-blind peer-review and meta-review process, we accepted 12 full papers and 10 short papers. To complement the oral presentations of short and full papers, the SLERD2017 program also included presentations of the best ideas from the 2017 international and local student contests (not included in these proceedings). These competitions challenged local and international students to propose ideas and proofs of concept/prototypes to make learning ecosystems smarter.

The selected scientific papers aim to understand, conceive, and promote innovative human-centric design and development methods, education/training practices, informal social learning, and citizen-driven policies. The papers are organized mirroring the main conference sessions in three themes, namely (i) the elaboration on the notion of smart learning ecosystems; (ii) the investigation of the relation of smart learning ecosystems with their territory; and (iii) the identification of resources for smart learning.

SLERD 2017 contributes to foster the social innovation sectors, identifying and discussing ICT and economic development and deployment strategies alongside with new policies for smarter proactive citizens. The proceedings are relevant for both researchers and policy makers.

In summary, SLERD2017 offered an exciting program that provided an excellent overview of the state of the art in smart learning ecosystems and was an occasion for bringing research forward and creating new networks.

We are very proud of the final selection of papers, which would not have been possible without the effort and support of our excellent Conference and Program Committees, including more than 50 international researchers. We would like to thank all the ones who, in different roles, have contributed their time to organize the event with enthusiasm and commitment.

April 2017   Monica Divitini
                                                                                  Óscar Mealha
                                                                                  Matthias Rehm

Conference Organization

General Chair

Fernando Ramos   University of Aveiro, Portugal

Conference Co-chairs

Carlo Giovannella   University of Rome Tor Vergata, Italy
Alke Martens   Universität Rostock, Germany

Honor Committee

Manuel Assunção   Rector of the University of Aveiro, Portugal
Ana Abrunhosa   President of CCDRC- Coordination Agency
                                     for the Portuguese Center Region, Portugal
José Ribau Esteves   President of CIRA-Aveiro Region
                                     Intermunicipality Agency, Portugal

Inter-Associations Committee

Darco Jansen   EADTU Programme Manager
Katherine Maillet   EATEL President
Radu Vasiu   IAFeS President
Airina Volungeviciene   EDEN President

Special Event Chairs

Giordano Bruno   ISIA Roma, Italy
Ana Veloso   University of Aveiro, Portugal
Annika Wolff   Open University, UK

Publishing Chairs

Monica Divitini   Norwegian University of Science
                                      and Technology, Norway
Óscar Mealha   University of Aveiro, Portugal
Matthias Rehm   Aalborg University, Denmark

Local Organizing Committee

Fernando Ramos   University of Aveiro, Portugal
Óscar Mealha   University of Aveiro, Portugal
João Batista   University of Aveiro, Portugal
Dora Pereira   University of Aveiro, Portugal
Rita Santos   University of Aveiro, Portugal
Cristina Silva   University of Aveiro, Portugal
Daniel Poças   University of Aveiro, Portugal

Scientific Committee

Ana Margarida   University of Aveiro, Portugal
   Pisco Almeida
Pedro Almeida   University of Aveiro, Portugal
Rui Aguiar   University of Aveiro/IT, Portugal
Marlene Amorim   University of Aveiro, Portugal
Diana Andone   Politehnica University of Timisoara, Romania
Maria João Antunes   University of Aveiro, Portugal
Vincenzo Baraniello   University of Rome Tor Vergata, Italy
João Paulo Barraca   University of Aveiro/IT, Portugal
João Batista   University of Aveiro, Portugal
Rosa Bottino   ITD – CN, Italy
Ilona Buchem   Beuth University of Applied Sciences
Bernardo Cardoso   Altice Labs, Portugal
Antonio Cartelli   Univ. of Cassino, Italy
John M. Carroll   Penn State University, USA
Ana Amélia Carvalho   University of Coimbra, Portugal

Pedro Carvalho — Altice Labs, Portugal
M. Paloma Diaz Perez — Universidad Carlos III de Madrid, Spain
Vincenzo Del Fatto — Bozen University, Italy
Giovanna Del Gobbo — University of Florence, Italy
Mihai Dascalu — Politehnica University of Bucuresti, Romania
Bertrand David — LIRIS – CNRS, France
Ines Di Loreto — Université de Technologies Troyes, France
Gabriella Dodero — Free University of Bolzano/Bozen, Italy
Jorge Ferraz de Abreu — University of Aveiro, Portugal
Davinia Hernandez-Leo — Universitat Pompeu Fabra, Spain
Marco Kalz — Open University, The Netherlands
Ralf Klamma — RWTH Aachen University, Germany
Milos Kravcik — RWTH Aachen University, Germany
Maria Beatrice Ligorio — University of Bari and CKBG, Italy
Pierpaolo Limone — University of Foggia and SIREM, Italy
António José Mendes — University of Coimbra, Portugal
João Filipe Matos — University of Lisbon, Portugal
José Mota — University of Aveiro, Portugal
Antonella Nuzzaci — University of L'Aquila, Italy
Lídia Oliveira — University of Aveiro, Portugal
António Osório — University of Minho, Portugal
Viktoria Pammer-Schindler — Know Center GmbH, Austria
Luís Pedro — University of Aveiro, Portugal
Dora Pereira — University of Aveiro, Portugal
Manuela Pinto — University of Oporto, Portugal
Elvira Popescu — University of Craiova, Romania
Rui Raposo — University of Aveiro, Portugal
Fernanda Ribeiro — University of Oporto, Portugal
Giuseppe Roccasalva — Politecnico Torino, Italy
Covadonga Rodrigo — UNED, Spain
Arnaldo Santos — Altice Labs, Portugal
Carlos Santos — University of Aveiro, Portugal
Rita Santos — University of Aveiro, Portugal
Marcus Specht — Open University, The Netherlands
António Teixeira — Universidade Aberta, Portugal
Filipe Teles — University of Aveiro, Portugal
Benedetto Todaro — Quasar Design University, Italy
Mário Vairinhos — University of Aveiro, Portugal
Giuliano Vivanet — University of Cagliari and SApIE, Italy
Imran Zualkernan — American University of Sharjah, United Arab Emirates

Additional Reviewers

Marius-Gabriel Guțu   University Politehnica of Bucharest, Romania
Ionel-Alexandru Hosu   University Politehnica of Bucharest, Romania

Voluntary Students

Carolina Abrantes   University of Aveiro, Portugal
Joana Beja   University of Aveiro, Portugal
João Jesus   University of Aveiro, Portugal
Diego Galego   University of Aveiro, Portugal
Elbênia Silva   University of Aveiro, Portugal
Tânia Ribeiro   University of Aveiro, Portugal

Contents

Smart Learning and Territory

Smart Learning Resources

Smart Learning Ecosystems

Mobile Seamless Learning Tool for Cancer Education

Nuno Ribeiro[1,2(✉)], Luís Moreira[3],
Ana Margarida Pisco Almeida[2], and Filipe Santos-Silva[1]

[1] Instituto de Investigação e Inovação em Saúde, Porto, Portugal
{nribeiro,fsilva}@ipatimup.pt
[2] Universidade de Aveiro, Aveiro, Portugal
{nunomgmribeiro,marga}@ua.pt
[3] Instituto Piaget, Vila Nova de Gaia, Portugal
luis.moreira@gaia.ipiaget.pt

Abstract. Mobile seamless learning provides the foundation for new digital solutions capable of adapting the contents to the learner needs and contexts. Studies have shown that cancer prevention knowledge remains reduced in modern societies. Given the worldwide burden of cancer, there is a need to increase cancer literacy in the populations, namely through the development of innovative strategies. This paper describes a 3 months two-arm quasi-experimental effectiveness study of a new mobile seamless learning tool for cancer education. Results showed that this application significantly increased cancer prevention knowledge of the users when compared to a control population ($p < 0.001$). This study provides evidence that a mobile seamless education tool can merge into the users' daily routine increasing users' knowledge by providing relevant cancer prevention information through messages delivered over an extended period of time. These innovative health education solutions will further expand the context of a smart learning ecosystem.

Keywords: Mobile learning · Seamless learning · Cancer prevention · Cancer literacy · Smartphone app

1 Introduction

A smart learning ecosystem can be described as a means to enable learning processes by removing or lowering barriers (Giovannella 2014). In this sense, mobile learning plays a very important role since it breaks down barriers and allows learning to take place in different contexts, enriching the learning processes (Park 2011). Mobile devices can be seen as "learning hubs", giving ubiquitous access to knowledge and connecting individual learners to communities of learners and physical or digital places. This is the underlying concept of mobile seamless learning, where a learner can learn in a variety of scenarios and contexts that can be easily switched given that it is mediated by a personal mobile device (Wong 2012). Mobile devices can make the educational process "just in time, just enough and just for me" (Peters 2007).

Smart cities are, in part, characterized by the emphasis on sustainable use of resources aimed at improving the well being of societies and the quality of life of its citizens

© Springer International Publishing AG 2018
Ó. Mealha et al. (eds.), *Citizen, Territory and Technologies: Smart Learning Contexts and Practices*,
Smart Innovation, Systems and Technologies 80, DOI 10.1007/978-3-319-61322-2_1

(Giovannella et al. 2012). On the other hand, cancer is set to become a major cause of morbidity and mortality in the coming decades in the world, thus representing an increasingly bigger burden on society (Bray et al. 2012). There is a need to develop smarter learning ecosystems regarding effective primary prevention strategies that might increase cancer awareness with corresponding healthier behaviours and early disease detection. Recent studies have shown some concerning results: general knowledge of certain cancer risk factors (such as alcohol, red and processed meats, low amounts of fruit and vegetables) and prevention behaviours remains alarmingly low (American Institute for Cancer Research 2017; Costa et al. 2016; Peacey et al. 2006; Sherman and Lane 2014). This calls for the need to better promote cancer literacy (Diviani and Schulz 2011) among the general public.

Smartphone applications can have a relevant role in these processes and have been pointed out as important tools to promote cancer prevention. A quick search in a major smartphone app distribution store such as Apple App Store or Google Play Store reveals several thousand health-related apps designed to promote smoking cessation, healthy eating, and other behaviours related with reduced risk of cancer. However, most of them lack scientifically validated data and haven't yet been tested in research studies to determine their effectiveness (Coughlin et al. 2016; Pandey et al. 2013).

The aim of this study was to assess the effectiveness of Happy, a cancer prevention smartphone app, as a mobile seamless learning tool for cancer education, which is based, among other features, on tailored messages adjusted to the users' context (location, time of day, week and month, weather conditions).

2   Methods

2.1   Happy, a Mobile Seamless Learning Tool for Cancer Education

Happy (Health Awareness and Prevention Personalized for You) is a cancer prevention smartphone app that aims to help users learn about cancer prevention in order to persuade them to make healthier choices, thus reducing their personal risk of developing several types of cancer.

Happy is based on the principle of tailoring, i.e., using information on a given individual/profile to determine what specific content he or she will receive (Hawkins et al. 2008). When users access Happy for the first time, they are required to answer a behaviour assessment questionnaire (Fig. 1a). The data collected allows the definition of the user profile and determines the current level of cancer prevention, called HappyScore (Fig. 1b). The HappyScore is represented on the landing page allowing the users to self-monitor their behaviour in a glanceable way.

Happy sends one tailored cancer prevention message a day to each user (Fig. 1c). This allows learning to occur through time in a non-intensive way, lowering the user burden and avoiding user rejection.

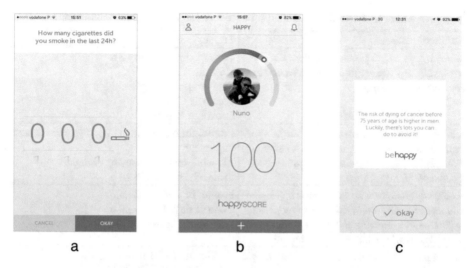

Fig. 1. Screenshots of the smartphone app Happy: (a) Sample question from the behaviour assessment questionnaire; (b) Landing page (HappyScore = 100); (c) Tailored cancer prevention message

The user profile is also used to tailor the cancer education messages sent to each user. Additionally, messages are tailored accordingly to the users' context (location, time of day, week and month, weather conditions). When a user is located in a target place, the message sent to her will take this in consideration, increasing the relevance and adequacy of the message content to the user (Table 1). The effort of tailoring messages to the users' profile and context is important because it allows the delivery of cancer prevention messages with less redundant information, that are more likely to be remembered and processed by the receiver (Hawkins et al. 2008). Also, it provides the needed adaptive flexibility to the app that allows mobile seamless education to occur. A total of 1,120 messages were developed. Messages follow the European Code Against Cancer guidelines (Schüz et al. 2015) and target specific risk factors of cancer, providing educational information.

Happy also allows behaviour monitoring. Users can track their behaviour by answering behaviour questions that are sent to them periodically by the app (1 each day) or by deliberately entering behaviour data. These behaviour assessments allow two different things in the context of the app. They are used to:

- update the user profile over time. This allows message tailoring to occur concurrently to the changes in behaviour emphasizing the risk factors that matter most to that user;
- recalculate the users' HappyScore. This shows the user the cause and effect link of a particular behaviour to the level of cancer prevention, an intense and very personal learning experience.

Happy also has a section that allows users to connect to each other, facilitating communication between them and a section with healthy challenges meant to engage the users with the app.

Table 1. Examples of tailored cancer education messages used in the app

Tailored message	Trigger context
Enjoying the beach? Beware: clouds don't protect from UV rays! Wear sunscreen and look for a shade. Be happy	Time of day: 12 h to 16 h UV index: >6 Temperature: >18°C Weather: Cloudy Location: Beach
Cervix cancer is the second deadliest form of cancer among young women. Don't risk your life. Book an appointment with your gynaecologist. Be happy	Profile: female, more than one year since last pap smear Time of week: Monday to Friday Time of day: 10 h to 20 h
At the supermarket? Did you know that by reducing the amount of salt you're reducing your stomach cancer risk? Avoid foods that come in a package like chips. Be happy	Profile: high salt consumption Time of day: 10 h to 20 h Location: Supermarket
Going out? Did you now that drinking alcohol raises your risk of cancer? Don't drink. If you do, stick to one drink a night. Be happy	Profile: alcohol drinker, female Time of day: 21 h to 23 h Time of week: Thursday, Friday or Saturday Location: Bar/Disco

2.2 Study Design

A two-arm quasi-experimental design was used with baseline (at study enrolment) and post-test (3 months later) assessments. All elements in the intervention group used the smartphone app Happy whereas the elements of the control group didn't. All app users (n = 3,252) received an email with a link to an online questionnaire in the beginning of the study. Users that answered the questionnaire and had more than 18 years were included in the intervention group. The same email was sent to a mailing list from University of Aveiro (n = 2,558). Every individual that answered the questionnaire, had a smartphone and had more than 18 years was included in the control group. Three months later, all study participants received a new email requiring them to answer the same online questionnaire a second time.

The online questionnaire consisted in two groups of questions:

- Sociodemographic characterization: age, gender, education, and email address;
- Knowledge assessment: 15 multiple choice questions concerning several dimensions of cancer prevention (risk factors, epidemiology and behaviour guidelines).

The primary outcome of the study was knowledge, measured as the difference between baseline and post-test knowledge assessments. The differences between baseline and post-test knowledge assessments were tested using independent and paired samples t tests (inter and intra groups, respectively).

3   Results

3.1   Participants

As shown in Fig. 2, a total of 3,252 and 2,558 participants were eligible for the intervention and control groups, respectively. Of these, 523 (16.1%) and 103 (4.0%) answered the baseline questionnaire. In the post-test assessment, 401 (76.7%) and 22 (21.3%) didn't answer the questionnaire and, thus, were lost to follow-up in the intervention and control groups respectively. One participant was excluded from the control group because she downloaded and used the intervention app.

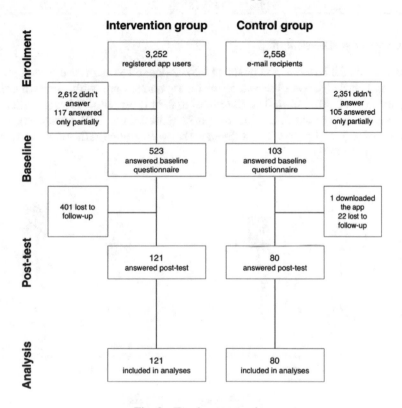

Fig. 2.   Enrolment overview

Participant characteristics are presented in Table 2. Participants were predominantly 31 to 40 years old (39.3%), female (77.6%), and had a college degree (86.1%).

Table 2. Demographic characteristics of study participants, n (%)

		All (n = 201)	Intervention group (n = 121)	Control group (n = 80)
Gender	Female	156 (77.6)	93 (76.9)	63 (78.8)
	Male	45 (22.4)	28 (23.1)	17 (21.3)
Age	18–25 years	73 (36.3)	33 (27.3)	40 (50.0)
	26–30 years	49 (24.4)	26 (21.5)	23 (28.8)
	31–40 years	79 (39.3)	62 (51.2)	17 (21.3)
Education	Undergraduate	28 (13.9)	16 (13.2)	12 (15.0)
	Graduate	173 (86.1)	105 (86.8)	68 (85.0)

3.2   Knowledge Assessment

Figure 3 shows the baseline and post-test knowledge assessment results. A significant increase in knowledge was observed on the intervention group, with a mean increase of 0.08 points (p < 0.001), from 0.66 (baseline) to 0.74 (post-test). Knowledge remained almost equal on the control group, from 0.70 (baseline) to 0.71 (post-test), a mean difference of only 0.01 points. The differences between groups in the post-test were also significant (p < 0.001).

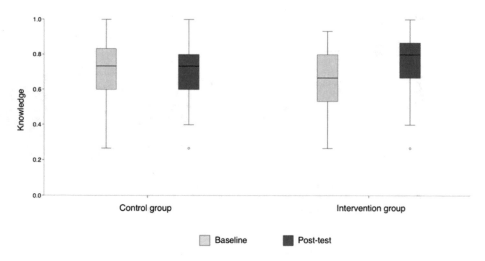

Fig. 3.   Baseline and post-test knowledge assessment results

4   Discussion and Conclusion

Smart learning ecosystems have the potential to revolutionize the way we look at education. Mobile devices, as "learning hubs", are at the core of this revolution. They provide the means to implement seamless learning systems, allowing learning to occur through numerous contexts and moments in time. This study aimed at assessing the effectiveness of Happy, a cancer prevention smartphone app, as a mobile seamless education tool for cancer education. Happy uses tailored cancer prevention messages as the main way to seamlessly educate users about cancer prevention. Users receive one message per day, tailored to the profile and physical context of the user. This allows the delivery of relevant cancer prevention messages that are more likely to be remembered and processed by the receiver. The app is designed to adapt to behaviour changes that might occur in users, guaranteeing that the content delivered is always relevant for the user.

The study was conducted with 201 participants distributed in two groups (intervention and control). Participants were mainly 31 to 40 years old, female and had a college degree. At baseline, knowledge assessment showed that both groups had a similar level of cancer prevention knowledge: 0.66 and 0.70 in the intervention and control group, respectively. The intervention group used the app for 3 months, whereas the control group didn't receive any intervention. The results of the post-test revealed that the level of cancer prevention knowledge remained the same in the control group and significantly increased in the intervention group ($p < 0.001$). This suggests that the app effectively contributed to changes in terms of cancer prevention knowledge. It's worth noting that this was a low burden intervention (users only received one message per day) and that the app was used in the context of everyday lives. Thus, this study provides evidence that a mobile seamless learning tool can be merged into the users' daily routine and still influence users' knowledge by providing relevant cancer prevention information through messages sent over an extended period of time (3 months). Happy was developed as a "just in time, just enough and just for me" cancer education tool. In this sense, it proved to be a good strategy to provide relevant cancer education content and, thus, increase cancer literacy in the studied population.

The proposed approach described in this article was based on a model that can be applied to different contexts, opening new possibilities in other educational themes or by integrating a larger smart learning ecosystem. More than an end solution to a problem, this approach should be seen as a means to provide continuity from formal to informal educational contexts, eliminating barriers in the learning process and contributing to create new modalities that enhance learning experiences in different domains and scenarios.

Acknowledgments. The authors would like to thank all the volunteers that participated in the study. This work was supported by the Gulbenkian Foundation through Project HYPE and by the Portuguese national funding agency for science, research and technology (FCT) [grant number SFRH/BD/92996/2013].

References

American Institute for Cancer Research (2017) 2017 AICR Cancer Risk Awareness Survey Report. https://vcloud.aicr.org/index.php/s/cehbAmj4IeQ8UDA?_ga=1.168227809.153246 2967.1486041236#pdfviewer
Bray F, Jemal A, Grey N, Ferlay J, Forman D (2012) Global cancer transitions according to the human development index (2008–2030): a population-based study. Lancet Oncol 13(8):790–801. doi:10.1016/S1470-2045(12)70211-5

Costa AR, Silva S, Moura-Ferreira P, Villaverde-Cabral M, Santos O, do Carmo I, Barros H, Lunet N (2016) Health-related knowledge of primary prevention of cancer in Portugal. Eur J Cancer Prev 25(1):50–53. doi:10.1097/CEJ.0000000000000125

Coughlin S, Thind H, Liu B, Champagne N, Jacobs M, Massey RI (2016) Mobile phone apps for preventing cancer through educational and behavioral interventions: state of the art and remaining challenges. JMIR mHealth uHealth 4(2):e69. doi:10.2196/mhealth.5361

Diviani N, Schulz PJ (2011) What should laypersons know about cancer? Towards an operational definition of cancer literacy. Patient Educ Couns 85(3):487–492. doi:10.1016/j.pec.2010.08.017

Giovannella C (2014) Smart Learning Eco-Systems: "fashion" or "beef"? J E-Learning Knowl Society 10(3):15–23

Giovannella, C., Iosue, A., Tancredi, A., Cicola, F., Camusi, A., Moggio, F., Baraniello, V., Carcone, S. and Coco, S.: Scenarios for active learning in smart territories. IxD&A **16**, 7–16 (2013)

Hawkins RP, Kreuter M, Resnicow K, Fishbein M, Dijkstra A (2008) Understanding tailoring in communicating about health. Health Educ Res 23(3):454–466. doi:10.1093/her/cyn004

Pandey A, Hasan S, Dubey D, Sarangi S (2013) Smartphone apps as a source of cancer information: changing trends in health information-seeking behavior. J Cancer Educ 28(1):138–142. doi:10.1007/s13187-012-0446-9

Park Y (2011) A pedagogical framework for mobile learning: categorizing educational applications of mobile technologies into four types. Int Rev Res Open Distance Learning 12(2):78–102

Peacey V, Steptoe A, Davídsdóttir S, Baban A, Wardle J (2006) Low levels of breast cancer risk awareness in young women: an international survey. European J Cancer (Oxf, Engl: 1990) 42(15):2585–2589. doi:10.1016/j.ejca.2006.03.017

Peters K (2007) m-Learning: Positioning educators for a mobile, connected future. Int J Res Open Distance Learning 8(2):1–17

Schüz J, Espina C, Villain P, Herrero R, Leon ME, Minozzi S et al (2015) European code against cancer 4th edition: 12 ways to reduce your cancer risk. Cancer Epidemiol 39:S1–S10. doi:10.1016/j.canep.2015.05.009

Sherman SM, Lane EL (2014) Awareness of risk factors for breast, lung and cervical cancer in a UK student population. J Cancer Education, 1–4. 10.1007/s13187-014-0770-3

Wong LH (2012) A learner-centric view of mobile seamless learning. Br J Edu Technol 43(1):E19–E23. doi:10.1111/j.1467-8535.2011.01245.x

Reflecting on Co-creating a Smart Learning Ecosystem for Adolescents with Congenital Brain Damage

Antonia L. Krummheuer[1], Matthias Rehm[2(✉)], Maja K.L. Lund[1],
Karen N. Nielsen[1], and Kasper Rodil[2]

[1] Faculty of Humanities, Aalborg University, 9000 Aalborg, Denmark
antonia@hum.aau.dk
[2] Technical Faculty of IT and Design, Aalborg University, 9000 Aalborg, Denmark
matthias@create.aau.dk

Abstract. Special needs education is focusing on a complex interplay of cognitive (knowledge), physical (motor rehabilitation), and social (interaction) learning. There is a strong discrepancy between the institutional spaces in which learning takes place and the need for scaffolding these levels of learning. In this paper, we present a first part of an ongoing collaboration with a special needs education facility for adolescents with congenital and acquired brain damage, that is interested in exploring the transformation of the institutional space into a smart learning ecosystem. We exemplify our research approach with a case study of a corridor in the institution that serves as a testbed for the involvement of all parties, i.e. residents, staff, management, in this transformation process.

Keywords: Smart learning ecosystem · Social practice · Co-creation · Brain damage

1 Introduction

One crucial aspect of smart learning ecosystems (SLEs) is their perspective change in terms of where education or learning takes place. The concept of SLEs allows for re-thinking traditional learning institutions tasks, creating room for informal and experiential learning and by doing so changing/modifying/adapting traditional learning approaches as well as the actual layout and design for the built environment. While this can be beneficial for all learners, we claim that it will be especially beneficial for learners that are challenged by the traditional educational system.

In this paper, we present a case study that investigates potentials for out-of-class learning for adolescents with congenital brain damage[1]. This study was done in the context of a long-term collaboration with a residency and rehabilitation center for adolescents (age 16–20) with moderate to serious brain injuries, both acquired and congenital. During their stay, adolescents participate in a three- to four-year educational program that is tailored to individual challenges and abilities and aims at improving

---

[1] The term *congenital brain injury* bundles various 'disorders' bound to a damage to the brain before, while of briefly after birth (Clemmensen-Madsen 2004).

© Springer International Publishing AG 2018
Ó. Mealha et al. (eds.), *Citizen, Territory and Technologies: Smart Learning Contexts and Practices*,
Smart Innovation, Systems and Technologies 80, DOI 10.1007/978-3-319-61322-2_2

cognitive, physical and social abilities. Thus, any attempt at transforming the institutional space into a smart learning ecosystem must tailor to at least one of these objectives, i.e. either contribute to cognitive development (e.g. conveying knowledge or reasoning skills), physical development (e.g. training motoric skills like moving an arm), or social skills (e.g. scaffolding social interaction or collaborative tasks).

In the center, each resident has its own apartment and is supported by an interdisciplinary team (therapists, pedagogues, social and health care workers). In discussions with staff and management a specific part of the building, a corridor, emerged as a space that seems to serve as an informal meeting place but does not encourage any interaction between residents. The decision to investigate this corridor also decided the potential group of users as the corridor is in the section for adolescents with congenital brain damage. All but one of the adolescents we observed were using a wheel chair, some of them could use it alone while others needed assistance. Some of the adolescents can talk, others use sounds, gestures, communication books or technologies steered with their hands or eyes for communication.

2   Related Work

We have argued above that SLEs might be especially beneficial for challenged learners and the realization of SLEs in traditional learning institutions might benefit this group. Schreiber-Barsch (2017) analyzes the role of space in relation to lifelong learning on the background of the UN Convention on the Rights of Persons with Disabilities and highlights the importance of the built environment for in- or excluding citizens from learning opportunities. This is also evident within disability studies that discuss the physical design of space as a crucial aspect of excluding, marginalizing and oppressing people with disabilities (Titchkosky 2011; Freund 2010; Imrie and Kumar 1998; Kitchin 1998; Hahn 1986). A prominent example for this organization is the segregation of people with disabilities to certain locations like schools or centers often outside or at the margins of the urban environment (which is also true for our collaboration partner).

Most studies touching on aspects of smart learning environments are though concerned with school or university class rooms, presumably due to intimate knowledge of the involved researchers with this context. The study presented by Jayasainan and Rekhraj (2015) reports on the potentials of scaffolding social engagement, informal learning, dialogue, and group work. The main advantage is seen in enabling learners to become stakeholders in their own learning process and thus assuming responsibility for their learning success. Of course, changing the space, itself is not automatically creating collaboration between learners, but it creates a place that encourages and supports a change in learning/pedagogical strategies (Divaharan et al. 2017). Divaharan and colleagues make it clear that such a change must be supported by the people engaged in the social practice of learning at the institution, i.e. teachers, learners and management alike.

Grigsby (2015) as well as Bilandzic and Foth (2014) analyze how a different traditional learning institution, i.e. the library, can change to become a smart learning environment as a hub of social learning and collaborative exploration of knowledge by embracing current technological trends, re-thinking the role of libraries, and re-designing the built environment of the library to cater to this development. On a more fundamental level shows Brooks (2011) the positive effect of technology-enhanced learning environments on learning outcome, highlighting the importance of the technical layer of smart learning environments.

In respect to the further transformation of the institution into a smart learning environment, Benze and Walter (2017) argue that involving citizens (in their case children and young people) into the planning activities of a given urban space will not only further learning about this space but has the chance to understand and take part in the intricate network of stakeholders involved in negotiating the future development of a space. This opens an interesting avenue of exploration for the overall process of transforming the whole institution into a smart learning environment. This also opens the question on who is going to conceptualize the space and its possibilities. Jornet and Jahreie (2013) describe that research usually focuses on analyzing user experience post-factum, i.e. when the design process has already resulted in a product. Instead they argue that it is worth looking at how the conceptualization of the space is negotiated in the design process, and by whom. With the example from designing a hybrid learning space for a museum, they show how the use of prototypes can become powerful tools for discussing the potentials of the space. Although they embrace the idea of analyzing the design process, they do not consider user involvement in this process.

As a side note, we are well aware of the discussion about the difference between space and place (e.g. Dourish 2006; Knox and Fincher 2013) but refrain from getting into this discussion here. Our concern is mainly with transforming the space inside the institution to enable user to engage in meaningful (learning) interactions, which will turn this space into places for the individual users. But those subjective interpretations of the space are not our concern.

3   Research Approach

Our research approach is in line with the approach described in depth in Rehm et al. (2016) for the specific case of developing social robots for institutional care settings.[2] In principle, we are an interdisciplinary team that is driven by the idea of developing technology together with users and stakeholders. We are specifically not user-centered but aim a co-creation of technology. We like to stress this point, because we have the feeling that current reports on user-centered design are often only marginally involving users. Thus, we embrace the idea of co-creation instead, where we rely on our specific users throughout the whole process. Because we work in institutional settings this means we cannot engage with single individuals but have to engage with a network of persons with diverse perspectives, from residents and their relatives over care personnel to management. Our goal is to identify social practices where the introduction of technology could make sense in the specific case, e.g. by increasing independence of residents or by freeing up time of staff, allowing them to concentrate on their core

---

[2] See also http://si.ehci.dk for an overview of projects and project partners.

competences. To this end, we build on a mix of methods from social sciences, humanities, and engineering allowing for gaining on the one hand a deep insight into the social practices surrounding the life of the residents in the institution and on the other hand a similar insight into the institutional rationales that will play a crucial part when introducing new technologies. Based on these insights we can develop targeted technological interventions that are based on real challenges residents and staff face in their daily life.

4   Analyzing Practices of Corridor Use

As mentioned in the introduction, we decided to work with the main corridor in one of the buildings. Figure 1 is a schematic drawing of the corridor highlighting the length as well as the different functions of the rooms that are located in this corridor, ranging from apartments over offices to therapy rooms. Figure 2 shows some impressions from the corridor, the left image taken at the entrance (corresponding to the left-most point of the drawing in Fig. 1, the middle one taken at the other end just in front of the common room, and the right one depicting an area for social interaction. This corridor was described by management and staff as unappealing and unwelcoming. At the same time, they called it a market place or pedestrian zone, clarifying that the corridor is not only a zone of transit but also a place for social interaction (Lu et al. 2011).

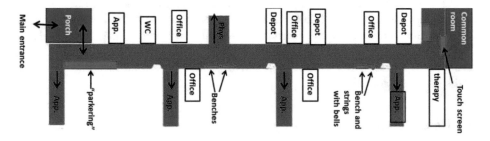

Fig. 1. Layout of the corridor

Fig. 2. Impressions from the corridor.

In order to get an insight into the use of the corridor we initiated a series of ethnographic observations to understand (i) the corridor, how it is used and what it affords, (ii) the users and their abilities as well as (iii) materials, activities and other actors on the corridor. Unlike traditional ethnography, our observations were clearly design-related, i.e. very focused with a clear goal, were done by several researchers, and had to be reached in a limited period of time (e.g. Crabtree 2012; Miller 2000; Hughes et al. 1994). Hughes and colleagues (1994) identify four different ways of adapting traditional ethnography to the need of design processes especially that of time pressure, relatively small scale of research focus and the non-interventionist role of the ethnographer. In our approach, we focus on what they call "concurrent ethnography", where the ethnographic study is taking place at the same time as the design process. This includes a close cooperation between ethnographer and designer informing each other during the process of fieldwork, debriefing, system design and prototype interaction, accompanied by ethnographic fieldwork which are repeated several times. These circles are combined with workshops and meetings with the target users, e.g. staff and residents, which also feed back into the development process.

Over the time of two months, 10 days were spent collecting data, either through observations, through design workshops or through in-situ interviews. In all activities, residents and staff participated. On the one hand, this allowed us to capture different perspectives, on the other hand this is often necessary because staff has to serve as interpreters between researchers and residents.

The observations in the corridor revealed that the corridor is indeed not only used as a space for transition but represents a space that is used for social interaction between resident. Summarizing from the observations, three distinct roles could be established that residents assume frequently on the corridor.

1. Looking for social contact: Residents use the corridor to get in contact with other residents or staff and seem to have a strong social awareness of others. This sometimes results in a kind of hunt to find an employee that has time to engage in short social interactions. At the same time, it became obvious that many have difficulties to initiate and/or maintain social interactions, esp. with other residents (see also Petry et al. 2005; Whitehouse et al. 2001; McWilliam and Bailey 1995).
2. Engaging in focused activities: The corridor has already been equipped with some technology that can be used for free-time activities like an area where strings with bells hang from the ceiling for sensual stimulation or a touch screen, where residents can listen to music. Mostly, these activities were done alone, but we observed one occasion where two adolescents were using the screen together and another occasion where two adolescents were dancing together on the floor. In the latter examples, it remains unclear whether these activities were joint activities or done side by side. Were the persons using the touch screen together, or was one person just watching the other using it? The employees reported that the adolescents in general had difficulties in engaging and undertaking joint activities with each other, which confirms our general impression.
3. Being a ratified bystander: Whereas in the above two categories, residents actively engage in either social interaction or focused interaction with objects, we could also observe residents that were seeking the vicinity of others but clearly did not wish to actively participate in the ongoing activities. The adolescents in this case claim the status of a "ratified bystander" (Goffman 1981) and employees (and other residents) readily accept their hybrid participation status as being close enough to be part of the group but not taking actively part in the joint activity. The employees described

these adolescents as spectators and compared the corridor to a pedestrian street in which people observe others.

5 From Institutional Space to Smart Learning Ecosystem

The observations were complemented with five workshops with residents and staff members and additional in-situ interviews with both residents and staff members.

At the first workshop, we discussed our observations with staff members to understand their perspective on the adolescents' practices on the corridor and to elicit aspects that should be considered for the design of the technology. Staff members wrote their comments and ideas on post-its which we took up for further discussion. In a second workshop, we engaged in a mutual understanding process with the adolescents. We presented a 3D-model of the corridor to engage a discussion of their understanding of the corridor with the possibility to enact certain scenarios. The third and fourth workshop were directed to initiate a creative process of developing concrete ideas of possible interactive technologies in the corridor and where to place them. The final event of this part of the project was a common lunch meeting with residents, staff, and management, where we concluded from the observations and workshops and pitched some first design ideas to the whole group. This resulted in lively discussions about the potential venues of the project. Especially the pedagogues and teachers, which had not been part of the workshops in this first part of the project could instantly see the potential of transforming the corridor into an (informal) learning environment and opted strongly for the possibility of dynamically relating the content to the curriculum, stressing again the three levels of learning identified earlier.

1. The design must consider different participation roles in a focused activity, as e.g. the ability to become a passive bystander observing activities from a close distance and allowing multiple users to use a technology at one time.
2. The design must ensure to keep the marketplace atmosphere, that is the possibilities of seeing, meeting and approaching people and activities.
3. The design must open up to multimodal ways of communication offering various ways of interacting with the content.
4. The design should take the three levels of cognitive, physical, and social learning into account.
5. The design should be considered as one building block in a smart learning ecosystem that encompasses the whole institution.

Based on these considerations, we initiated the next iteration of this project, where we are currently concentrating on developing several prototype installations together with residents and staff that will allow for more focused discussions on the technical possibilities.

6 Future Work

In this paper, we presented the first step in our research on the role of space and place in institutional learning contexts, where we analyzed in depth the use of a space that is prominent in the life of the residents but is so far only used (and seen) as a non-place, a place of transition from one meaningful space in the institution to another. The workshops with the residents and staff revealed the potential for changing the meaning of this space and turning it into a place for (informal) learning.

The corridor is just one specific area in the building and one can easily imagine an interactive installation inside the corridor. But especially the last workshop opened up to the possibilities of turning the whole institution into a smart learning ecosystem, and thus raised more questions than it answered, e.g. What are the features in relation to this specific group of learners and teachers? Teaching at the institution is based on individual curricula depending on abilities of the resident. How can that be reflected in casual collaborative encounters outside the classroom? How can the classroom learning goals feed into the informal and experiential learning throughout the built environment of the institution? Are there any formal models for this relation/mapping? How has this built environment to change to enable the residents in their learning?

Acknowledgements. We would like to thank all citizens, staff members, and management at Neurocenter Østerskoven for their cooperation and dedication in this project.

References

Benze A, Walter U (2017) The neighbourhood as a place of learning for young people. Springer International Publishing, Cham, pp 147–158. http://dx.doi.org/10.1007/978-3-319-38999-8_14

Bilandzic M, Foth M (2014) Learning beyond books|strategies for ambient media to improve libraries and collaboration spaces as interfaces for social learning. Multimedia Tools Appl 71(1):77–95. http://dx.doi.org/10.1007/s11042-013-1432-x

Brooks DC (2011) Space matters: the impact of formal learning environments on student learning. Br J Educ Technol 42(5):719–726. http://dx.doi.org/10.1111/j.1467-8535.2010.01098.x

Clemmensen-Madsen T (2004) Ny indsigt - ny indsats: udviklingsprojekt til intensivering af optræningsindsatsen for børn med medfødt hjerneskade. MarselisborgCentret, Århus

Crabtree A, Rouncefield M, Tolmie P (2012) Doing Design Ethnography. Springer, London

Divaharan S, Wong P, Tan A (2017) NIE Learning space: physical and virtual learning environment. Springer, Singapore, pp 253–265. http://dx.doi.org/10.1007/978-981-10-3386-5_14

Dourish P (2006) Re-space-ing place: "place" and "space" ten years on. In: CSCW 2006: Proceedings of the 2006 20th anniversary conference on computer supported cooperative work. ACM, New York, pp 299–308 (2006)

Freund P (2010) Bodies, disability and spaces: the social model and disabling spatialorganisation. Disabil Soc 16(5):689–706

Goffman E (1981) Footing. In: Goffman E (ed) Forms of talk. Blackwell, Oxford, pp 124–159

Grigsby SKS (2015) Re-imagining the 21st century school library: from storage spaceto active learning space. TechTrends 59(3):103–106. http://dx.doi.org/10.1007/s11528-015-0859-5

Hahn H (1986) Disability and the urban environment: a perspective on Los Angeles. Environ Plan D Soc Space 4:273–288

Hughes J, King V, Rodden T, Andersen H (1994) Moving out from the controlroom: ethnography in system design. In: Proceedings of the 1994 ACM conference on computer supported cooperative work. ACM Press (1994)

Imrie R, Kumar M (1998) Focusing on disability and access in the build environment. Disabil Soc 13(3):357–374

Jayasainan SY, Rekhraj HS (2015) X-Space: AWay forward? The Perception of Taylor's University Students on collaborative learning spaces. Springer, Singapore, pp 411–429. http:// dx.doi.org/10.1007/978-981-287-399-6

Jornet A, Jahreie, CF (2013) Designing for hybrid learning environments in a science museum: inter-professional conceptualisations of space. In: Childs M, Peachey A (eds) Understanding learning in virtual worlds. Springer, London, pp 41–63

Kitchin R (1998) 'Out of Place', 'Knowing One's Place': space, power and the exclusion of disabled people. Disabil. Soc 13(3):343–356

Knox D, Fincher S (2013) Why does place matter? In: Proceedings of the 18th ACM conference on innovation and technology in computer science education, ITiCSE 2013. ACM, New York, pp 171–176. http://doi.acm.org/10.1145/2462476.2465595

Lu Z, Rodiek SD, Shepley MM, Duffy M (2011) Influences of physical environment on corridor walking among assisted living residents findings from focus group discussions. J Appl Gerontol 30(4):463–484

McWilliam RA, Bailey DB (1995) Effects of classroom social structure and disability on engagement. Top Early Child Spec Educ 15(2):123–147

Miller DR (2000) Rapid ethnography: time deepening strategies for hci field research. In: Boyarski D, Kellogg WA (eds.) Proceedings of the 3rd conference on Designing interactive systems: processes, practices, methods, and techniques. ACM, New York (2000)

Petry K, Maes B, Vlaskamp C (2005) Domains of quality life of people with profound multiple disabilities: the perspective of parents and direct support staff. J Appl Res Intellect Disabil 18(1):35–46

Rehm M, Krummheuer AL, Rodil K, Nguyen M, Thorlacius B (2016) From social practices to social robots: user-driven robot development in elder care. Springer International Publishing, Cham, pp 692–701. http://dx.doi.org/10.1007/978-3-319-47437-3_68

Schreiber-Barsch S (2017) Space is more than place: the urban context as contested terrain of inclusive learning settings for adults and arena of political subjectivation. Springer International Publishing, Cham, pp 67–81

Titchkosky T (2011) The Question of Access: Disability, Space, Meaning. University of Toronto Press, Toronto

Whitehouse R, Chamberlain P, O'Brian A (2001) Increasing social interactions forpeople with more severe learning disabilities who have difficulty developing personal relationships. J Intell Disabil 5(3):209–220

Creativity Enhanced by Technological Mediation in Exploratory Mathematical Contexts

Artur Coelho[✉] and Isabel Cabrita

Research Centre for Didactics and Technology in the Education of Trainers (CIDTFF),
Dep. of Education and Psychology, Universidade de Aveiro, Aveiro, Portugal
{artur.coelho,icabrita}@ua.pt

Abstract. Creativity is fundamental to the sustainable development of societies. However, lack a School model that promotes it. In Mathematics, exploratory open-ended and challenging tasks, based on the effective resolution of and well oriented confrontation and discussion moments are needed to foster this skill. The digital revolution brought a set of technological tools with great potential in the educational context, particularly to engineer collaborative work environments and to mediate communication. But, the use of these tools remains inadequate.

The main objective of this qualitative study was to evaluate the potential of digital technologies to construct collaborative environments and as communication mediation tools and how these dynamics influence the development of creativity and communication in Mathematics as well as students' (10–11 years old) digital literacy.

Preliminary results suggest that the implementation of these technologies allow to develop cross-curricular and specific mathematical skills and digital literacies, and truly change the current educational paradigm.

Keywords: Creativity · Technological mediation · Mathematics · Exploratory tasks

1   Introduction

The digital revolution is transforming the way we work, think, and "connect" with profound implications for cultural values (Castells 2007; Lévy 2010) and giving rise to new important societal challenges. We live in times of great uncertainty and unpredictability and one purpose of the "School" as an institution is to develop in its students an understanding of this fluid and changing world they live in. Ensuring a sustainable future for the next generations in this context requires a strong focus on innovation and creativity. However, our students have been taught and trained in mechanized procedures (Robinson 2011). Creativity, as a transversal skill to all areas of knowledge, has thus

This article reports research developed within the PhD Program Technology Enhanced Learning and Societal Challenges, funded by Fundação para a Ciência e Tecnologia, FCT I. P. – Portugal, under contract # PD/00173/2014 and # PD/BI.

© Springer International Publishing AG 2018
Ó. Mealha et al. (eds.), *Citizen, Territory and Technologies: Smart Learning Contexts and Practices*,
Smart Innovation, Systems and Technologies 80, DOI 10.1007/978-3-319-61322-2_3

been systematically curtailed by the educational systems of the industrialized world (Robinson and Aronica 2009). Several investigations (Franke et al. 2007) confirm that Maths continues to be taught in the traditional routine way where moments of mechanized practice follow the instructional exposition and subjects are explored in a disconnected way, isolated from other disciplinary areas and everyday life (Franke et al. 2007). To counter this practice, and given the demands of our society, students should be given opportunities to undertake challenging mathematical tasks that promote mathematical reasoning, communication, and creativity, that flourishes from confrontation and collective discussion of exploratory learning tasks.

Digital communication technologies have been widely adopted by European educational institutions: 96% of schools have access to the Internet and 80% of teachers say they recognize advantages in using computers (Korte and Hüsing 2006). But they are used especially for administrative and management purposes and there were no significant pedagogical changes, namely in terms of teaching strategies, classroom resources and students' learning (Punie et al. 2006). In fact, pedagogical approaches, with decisive impact on students' learning, are rarely creative (Redecker et al. 2009) and technology remains, mostly, to support direct teaching techniques. Thus, understanding the relationship between technology and the development of creativity in Maths in exploratory learning contexts was the main question for this study. The main intention was to evaluate the potential of these digital technologies in the construction of collaborative learning environments and as mediators of communication and how these dynamics influence the development of creativity in problem solving, mathematical communication and, simultaneously, of technological skills. Thus, we sought to contribute to the multimedia in education field. In particular, we aimed to better understand the contexts and phenomena associated with Maths when mediated by technology, and how this relationship determines, in the early years (10–11 years old), the development of emerging skills such as creativity (Coelho and Cabrita 2017).

2   Theoretical Framework

Common people still see creativity as the product of talent and personality traits (Amabile and Pillmer 2012). Studies in the 1960s and 1970s (Davis and Manske 1968; Speller and Schumacher 1975) reinforced the idea that creativity was associated with innate individual talent of a few exceptional individuals and could not be learned and taught. But Silver (1997) states that creativity is associated with a deep and flexible knowledge, which implies hard work and reflection and, consequently, should not be focused on individual exceptionality but understood as a phenomenon capable of being developed through education and training. In creativity, novelty is a central aspect, but it is not a sufficient feature, products need to be relevant, effective, ethical and moral concerns should be attended to (Cropley 2011; Runco and Jaeger 2012). So, creativity can be defined as a social phenomenon that observes specific rules and can be promoted or inhibited through social factors (Cropley 2011). In Maths, Gontijo (2007) understands creativity as the ability to find differentiated (unusual) ways of solving problems and finding original resolutions for non-traditional problems.

Several authors (Alencar 1990; Leikin and Pitta-Pantazi 2013; Silver 1997) consider that creativity in Maths is characterized by several dimensions, which can be measured through: fluency − the quantity of different ideas produced on the same subject; flexibility − the number of responses of distinct categories, which reflects the capacity to change reasoning; originality − the ratio between the frequency of an infrequent or unusual response and the number of students observed; elaboration − the amount of details of a given idea and it can be measured by their quantity. Almost two decades have passed since Cropley wrote that conventional education often hampers the development of the skills, attitudes and motivation necessary for innovation and, among other things, it perpetuates the idea that there is always only one correct answer to each problem. In this context, students only acquire skills that are necessary to produce orthodoxy (Cropley 2009). It is astounding, yet at the same time disturbing, that this vision remains so current/prevalent.

The development of higher mathematical skills such as creativity is not compatible with low complexity tasks mainly focused in a specific mathematical topic. Rather, it requires challenging tasks capable of stimulate intra-mathematical, other areas and day-to-day connections. These tasks allow students to approach Maths in a more realistic and positive way and encourage more diverse approaches. After their effective resolution by the students, it is critical to discuss them collectively (Stein, Engle, Smith, and Hughes 2008), aspect that deserves special attention in this ongoing research. So, teacher's role must evolve from an "instructor" and "guardian of the mathematical correction" to the "engineer" of these learning environments (Stein et al. 2008).

Being *connected* is a characteristic feature of the Knowledge Society, but knowing when, what and how to connect is a critical meta-skill (Castells 2007; Downes 2012). Siemens (2004) proposes Connectivism, as a learning theory for the digital age, based on the creation, maintenance and use of network connections (nodes) by its users and whose relations arise from their common interests. Nowadays, these interactions, markedly mediated by technology, are amplified by the enormous potential of the multidirectional communication established between individuals and institutions, and constitute true virtual communities. But some studies point out the difficulty of controlling students' activities in lessons, such as Maths, which take place in computerized laboratories or by using portable devices with Internet access (Galluch and Thatcher 2011; Tarafdar et al. 2013). Classroom Management Software [CMS], such as iTALC, allows the effective performance of these functions within the digital classroom. And Microsoft Office 365 contains a set of applications and services integrated in the cloud (cloud computing) that allows the development and management of activities in technological contexts. These applications can help to create integrated learning environments with features like file/resource sharing and synchronous and asynchronous interaction between participants, for example to launch topics and discussions in forums and groups, provide content in different ways − text, video or audio − or to track and monitor students' work at home, in several platforms and operating systems and at multiple scales. Although these digital resources are widespread in schools, they rarely correspond to significant changes in pedagogical strategies. Hence, in these new educational scenarios, new approaches are required to enhance their advantages while minimizing any constraints. These strategies must develop more favorable attitudes towards Maths and essential skills such as creativity and critical thinking and foster a better understanding of mathematical concepts involved. Digital tools can offer a spatial and temporal expansion of the classroom, reducing, in certain contexts, costs for students. This can lead to a growing diversification of the school population, more personalized learning and develops students' autonomy (Farnsworth et al. 2015).

There are several ways of combining face-to-face and distance activities. The Flipped Classroom (Bergmann and Sams 2014) is a model that reverses the traditional instructional process. A distance education strategy is adopted, where teachers are responsible for making assignments with a careful choice of resources – audio and video casts, links to specific content and open educational resources – that are delivered, in a digital platform, to individuals. Then follows the presential moment with the discussion and exploration of the presented tasks carefully guided by the teacher (Bergmann and Sams 2014).

3 Methodology

The qualitative case study selected (Bogdan and Biklen 1994) focused on three groups of two 6th grade students (11 years old) to allow an in-depth study. Their selection and distribution considered their mathematical performance – low, medium and high –, their expectations regarding Maths (only the most performant group saw utility in the discipline) and they all attended to every moment of the instructional sequences performed in a learning Math lab. Every session was a complement to formal Maths classes and were developed in a context close to them but the frequency was voluntary. The researcher had an active participation since he conducted all the events resulting from this research. The data were collected through several techniques: participant observation carried out by the teacher/researcher, supported by field notes and Logbook; survey, through questionnaires and interviews with the case students and a documentary analysis of students' task resolutions, Initial and Final tests and official documents produced by the school. First, we applied an Initial Questionnaire (IQ) and a ICT test to obtain information about student habits and basic knowledge of digital tools and applications, including Dynamic Mathematics Software [DMS] and social networks. This allowed us to calibrate the exploratory tasks in the two instructional sequences. An Initial Test (IT) in the beginning of each, solved with paper and pencil, checked previous students' knowledge on several mathematical topics and advised adjustments in future tasks. It also allowed us to assess students' improvement, when results were compared with the same Final Tests (FT) at the end of each sequence. Throughout this initial stage, the customizable and secure cloud computing Office 365 platform was used. Mainly, was used social features, like Yammer, allowing the creation of groups with different levels of interaction and access; synchronous and asynchronous communication tools – Outlook Mail, Outlook Groups and Skype; the streaming application Videos, to broadcast audio and video content; OneNote, as the digital daily notebook when working with computers, tablets and smartphones, on iOS, Android and Windows platforms. We implemented two didactic sequences in several sessions throughout the year, each consisting in a set of exploratory tasks with increasing complexity, both mathematical and technical. Task resolution was supported by digital tools – Interactive Whiteboards, Roamer robots, DMS, visual programming environments, namely KODU – and measuring instruments, manipulatives and paper and pencil. The physical classroom environment was mediated by iTALC, which allows to monitor students' efforts, make simultaneous demonstrations on their desktops and remotely control their terminals. It also allows students to share, on the whiteboard, their own desktop, which was particularly important in the presentation and discussion stage of task resolution. The teacher's desktop was constantly projected in the interactive whiteboard to support real-time interaction with every computer/tablet in the classroom (see Fig. 1).

Fig. 1. The teacher's desktop projected *(left)*. iTALC with nine active workstations *(right)*.

The first didactic sequence was implemented at the beginning of the second trimester and was divided into three parts: the first one was aimed to develop basic skills to operate a Roamer robot in numerical, geometric and algebraic contexts, namely by exploring circuits and measuring and proportions and ratios through scales. The second part contained more open-ended tasks, using the Roamer to draw increasingly complex flat figures involving concepts related to angles, lines, perimeters and areas, measuring, spatial and proportional reasoning. The third part contained a set of programming tasks in Kodu. They were aimed to develop logical reasoning and to provide a strong understanding of the programming environment that would be critical to solve more advanced tasks. These included the construction of small two-dimensional geometric scenarios (involving areas and perimeters of flat shapes) and more open and complex three-dimensional "worlds" with advanced notions of volumes (see Fig. 2).

Fig. 2. 3D world *(left)* created by students in KODU and visual code lines *(right)*.

The second didactic sequence was implemented in the third and last trimester. The resolution of exploratory activities, with several open-ended situations, required instruments, paper and pencils and the DMS GeoGebra, while exploring (compositions of) isometries – translation, reflection and rotation – and the concept of symmetry associated with the last two transformations. Prior to physical classroom sessions, we provided critical content in different formats – video, audio and text – through the Office 365 services and applications.

Students reach the classroom sessions with good knowledge of the subjects, instruments and tasks to be carried out, having already begun discussions and previous analysis on the group forum (Yammer) and in OneNote digital notebook, which allows "handwritten" annotations (using a digital pen on the tablet) and which can synchronize in real-time. Thus, the face-to-face moment in the classroom was greatly optimized because students had a prior contact with the main concepts. So, the sessions were dedicated to solving and discussing more open and complex tasks, which contributed to a more solid mathematical conceptualization and to the development of mathematical competence. The presential moment was implemented in four different stages (Stein et al. 2008). In the first stage, the task was orally presented, and some relevant aspects were clarified by the teacher. In the next stage, all groups solved their tasks autonomously but under teacher's supervision through the CMS or OneNote when tasks were performed in the computer/tablet. In the third stage, the working groups presented their resolutions. Computer tasks were assisted by the CMS features. Finally, students drew conclusions writing short reports on the digital notebook. Every group's resolutions were discussed by the whole class and everybody took note of the main ideas. At the end of each session, we "collected" students' work and analyzed the field notes to improve the Logbook. All these documents were assessed before the next session, so that the plan could be changed, if necessary. We intended to create an environment where students could make mistakes, with time to reason at their own pace and with opportunities to discuss and share their own ideas with everyone. This environment, physical and virtual, was a "place" of confrontation and discussion where technology was used mainly as a social collaborative learning and mediation tool. An open "classroom" beyond physical walls and the constraints of institutional schedules, capable of strengthening a collaborative working culture.

After the implementation of the two didactic sequences, we used a Final Questionnaire (FQ) and conducted several interviews with the selected group students. The Final Questionnaire aimed to collect data on their opinion about the entire project. The IT and FT, in each learning sequence, had a double purpose: the IT gave us information on the pupil's knowledge and skills before the didactic sequence implementation and the FT allowed us to assess their progress throughout the study.

All collected quantifiable data are under statistical analysis using Excel and will be presented through tables and graphs. Qualitative data are being subjected to content analysis through qualitative analysis software, using categories related to: Geometry – Angles and lines; spatial reasoning; perimeters, area and volume; measuring and; isometries and symmetry; Proportional reasoning and dimensions of Creativity.

4   Results

A preliminary statistical and qualitative analysis, ongoing, of the collected data allow us to present some previous results. Direct observation and the analysis of students' answers to the FQ and to the interview show the importance they gave to the structure of the learning space/environment, to the technological nature of the task, its exploratory and open nature and how they were addressed and discussed in the classroom as well as their contribution to the development of their creativity. The comparative analysis of the IT and FT made it possible to verify that the students with lower performance in Maths seemed to benefit more from the technological nature of the environment and the tasks (see Table 1).

Table 1.   Initial and final tests results (%).

		IT 1	FT 1	IT 2	FT 2
Group 1	Pupil a	28	46	36	51
	Pupil b	35	53	38	58
Group 2	Pupil c	61	68	72	79
	Pupil d	52	60	57	68
Group 3	Pupil e	86	91	87	95
	Pupil f	84	92	86	91

Most students stated in the interview that the proposed situations were familiar and exciting to them, and made sense of the mathematical contents. They further mentioned that the availability of content in various formats accessible through different platforms was very useful and versatile. They appreciated the prior contact (in Videos and OneNote) with the content since they could study and prepare the future session tasks. They also reported that synchronous (through Yammer and Skype) and asynchronous interactions (through Yammer, Outlook Groups and Email) with their colleagues and/or teacher helped them to clarify some doubts and to see other perspectives. At the beginning, students were unaware these applications but learned how to use them quickly and spontaneously – *"Several students began using the Yammer's synchronous communication tool to communicate between them and to ask me questions about the tasks. They did it from home computers and smartphones"* (Logbook entry, 06/01/2016). In these interactions, some students spontaneously assumed leadership roles – *"Two students asked to have admin rights in the Yammer forum. They are natural leaders and we gave them that role"* (Logbook entry, 14/01/2016). All students especially enjoyed the OneNote "digital notebook", wherewith they could receive real-time feedback, correct the resolutions, and complete assignments, anytime, anywhere, from home and in different devices, namely smartphones (see Fig. 3). But many mentioned that some software, such as instant messaging applications or social networks notifications, interfered negatively in their work, especially when it was accomplished at home. They added that they found it difficult to repress the curiosity and willingness to respond to their requests (Shirky 2014).

Fig. 3. Students solving a task in OneNote with real time feedback *(handwritten notes).*

Regarding creativity, we could also observe that the best performing students in Maths did not always look for alternative or improved resolutions to the tasks, and often was satisfied with a simply correct one. This also suggests that the exploratory open-ended nature of the tasks while having the potential to generate more creative resolutions (Fig. 4), needs an appropriate attitude to produce results.

Fig. 4. To solve task Ia, 3students need to program the Roamer to draw the line on the left. This group use very peculiar way - every single instruction is reversed - but the solution is correct.

On this matter, two students stated in the interview that *"most teachers appreciate that the students respond to the questions as quickly and in as much agreement as possible with what they had been taught"* (15/06/2016). And that the factors that most inhibited them from freely participating in collective discussions were the criticism of colleagues and the overly sanctioning and, sometimes, aggressive attitude of some teachers. This is not due to technology. It is an attitudinal matter that the school must counteract. On the other hand, the group of students with better performance initially revealed some discomfort in sharing their resolutions. This attitude combined with a highly competitive spirit has proved an obstacle to the establishment of a true collaborative work culture, with impacts on the overall creativity. In contrast, it was observed that the group of students less "orthodox" in task resolution produced less calculated but riskier interventions, facing error in a more natural and relaxed way and tended to present more creative products – *"These group of students tried to solve the maze problem at least 5 times. Small but critical mistakes didn't draw them back. They still excited with the perspective of success."* (Logbook entry, 04/02/2016). Supporting this idea, several students answered, in the FQ, that they "lost their fear of making mistakes" realizing that trial and error strategies were part of the process (see Fig. 5). Several students also mentioned that when they observed the creative work of their colleagues and teacher's appreciation of it, they felt compelled to improve theirs — *"Original solutions tend to trigger strong reactions. Other students felt motivated and committed to improve their own work."*(Logbook entry, 20/04/2016).

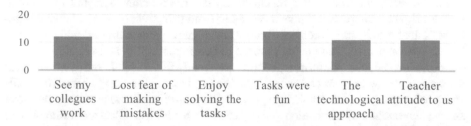

Fig. 5. Main aspects to foster creativity (number of students). Source: FQ.

All students declared, in the FQ and/or in the interview, that iTALC encouraged teamwork at classroom level. 80% of them also considered that this software was very important to manage the whole process of resolution, presentation and discussion, because it allowed the teacher to properly control all workstations. These aspects were highly valued by 85% of the students who noted the ease of interacting with the rest of the class and reported that knowing that the system was active kept them more focused on solving tasks. 90% of the students said that they *"strongly disagreed with the idea that this software's purpose was for controlling them."* It was also observable that the most fruitful discussions appeared when the students actually shared their ideas, and so they increased in number and quality throughout the study. And the increasing use of applications (even initially unknown) in specific contexts was observable (Fig. 6).

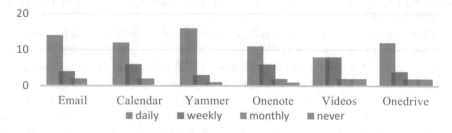

Fig. 6. The frequency of use of Office 365 applications (number of students). Source: FQ.

There was also an increase in the diversity of platforms used, in the access places and in the quality of the interactions, especially in OneNote. These facts support the perception of a sustained development of digital literacy in students.

5   Final Remarks

The research undertaken allows us to conclude, in a preliminary way, that the proper use of Office 365 Cloud Computing applications has several advantages that should be harnessed and enhanced. Gains in effectiveness in communication and interaction, through Email, Groups and Yammer, were high and using OneNote as a "digital notebook," capable of hosting a content library, task sequence protocols, and students' resolutions with real-time "hand notes", proved to be of great utility and versatility. These multi-system and multiplatform applications can really foster collaborative work within the class group, with other groups and with the teacher, anytime, anywhere. The availability of content, in different formats, in a digital platform and the use of synchronous and asynchronous communication tools together with a Flipped Classroom strategy can be used to build personalized learning environments not only of an intrinsic collaborative nature, but also very flexible and customizable, able to meet the specific needs of different students as stated by Bergmann and Sams (2014). The use of CMS in a classroom can foster dynamics interaction in a technological environment, providing opportunities to easily share ideas, to collaborate and to successfully support the discussion and confrontation moments, which reinforces results obtained previously by Coelho and Cabrita (2015). Associated with specific network filters, they also prevent students from diverting their attention to potentially disruptive requests, especially by social networks or instant messaging applications, thus keeping them focused and engaged with task resolution. Genuine interactions and sharing in a learning community are the core of a true collaborative working culture and, simultaneously, reinforce the sense of belonging and help to prevent the alienation of less proficient students, results that corroborate the defended by Farnsworth et al. (2015). These collaborative environments seem to increase students' autonomy and their levels of confidence, motivation and engagement, which implies a greater involvement in school life and a more favorable and interested attitude towards Maths.

We also concluded that implementing exploratory open-ended tasks in technological contexts seem to contextualize Maths, which becomes more clear, appealing and useful to the students. They felt challenged and motivated by these tasks and they worked hard to mobilize many mathematical concepts to solve them that are especially familiar to them when linked to technology. This aspect not only helped them to better understand Maths but also facilitated the emergence of more creative productions, collective and individual ones that grow from large group discussions and interactions, virtual and physical as stated by Levenson (2013). Students used different approaches when solving these tasks, and different procedures to find solutions, sometimes unique and with enriching details.

Therefore, there have been increases in various dimensions of creativity such as originality, fluency, flexibility and elaboration.

The study also indicates an effective and contextualized increase of students' digital literacies. Some of these aspects should be the object of much more extensive and detailed studies.

Their relevant role in teaching and learning Maths should have implications in teacher training.

References

Alencar EM (1990) Como desenvolver o potencial criador: um guia para a liberação da criatividade em sala de aula. Vozes, Petrópolis

Amabile TM, Pillemer J (2012) Perspectives on the social psychology of creativity. J Creative Behav 46(1):3–15. doi:10.1002/jocb.001

Bergmann J, Sams A (2014) Flipped Learning: Maximizing Face Time. T&D 68(2):28–31 http://search.ebscohost.com/login.aspx?direct=true&db=bch&AN=94004879&site=ehost-live

Bogdan RC, Biklen SK (1994) Investigação qualitativa em educação: uma introdução à teoria e aos métodos. Porto Editora, Porto, p 336

Castells M (2007) A Galáxia Internet, Reflexões sobre a Internet, Negócios e Sociedade, 2a edn Fundação Calouste Gulbenkian, Lisboa

Coelho A, Cabrita I (2015) A creative approach to isometries integrating GeoGebra and Italc with paper and pencil environments. J Eur Teach Educ Netw (JETEN) 10:71–85

Coelho A, Cabrita I (2017) Mediação tecnológica e desenvolvimento da criatividade em contextos matemáticos exploratórios. In: Submetido ao VIII Congreso Iberoamericano de Educación Matemática (VIII CIBEM) (p submitted). Madrid

Cropley AJ (2009) Creativity in education and learning: a guide for teachers and educators. Routledge Falmer, New York

Cropley AJ (2011) Definitions of creativity. In: Runco MA, Pritzker SR (eds) Encyclopedia of Creativity, 2nd Editio, pp 358–368. Elsevier. http://doi.org/10.1016/B978-0-12-375038-9.00066-2

Davis GA, Manske ME (1968) Effects of prior serial learning of solution words upon anagram problem solving: a serial position effect. J Exp Psychol 77(1):101–104. doi:10.1037/h0025791

Downes S (2012) Connectivism and Connective Knowledge: essays on meaning and learning networks. National Research Council Canada. http://scholar.google.com/scholar?hl=en&btnG=Search&q=intitle:Connectivism+and+Connective+Knowledge+Essays+on+meaning+and+learning+networks#0

Farnsworth V, Kleanthous I, Wenger-Trayner E (2015) Communities of practice as a social theory of learning: a conversation with etienne wenger. Br J Educ Stud 1005, 0–22. http://doi.org/10.1080/00071005.2015.1133799

Franke KL, Kazemi E, Battey D (2007) Mathematics teaching and classroom practice. In: Lester FK (ed) Second handbook of research on mathematics teaching and learning. Information Age Publishing, Charlotte, pp 225–356

Galluch PS, Thatcher J (2011) Maladaptive vs faithful use of internet applications in the classroom: an empirical examination. J Inf Technol Theory Appl 12(1):5–22

Gontijo CH (2007) Estratégias de ensino em matemática e em ciências que promovem a criatividade. Ciência & Ensino 1(2):10

Korte WB, Hüsing T (2006) Benchmarking access and use of ICT in European schools 2006: results from head teacher and A classroom teacher surveys in 27 European countries. In: Current Developments in Technology-Assisted Education

Leikin R, Pitta-Pantazi D (2013) Creativity and mathematics education: the state of the art. ZDM 45(2):159–166. doi:10.1007/s11858-012-0459-1

Levenson E (2013) Tasks that may occasion mathematical creativity: teachers' choices. J Mathe Teacher Educ 16(4):269–291. doi:10.1007/s10857-012-9229-9

Lévy P (2010) Cibercultura, 3ª edn. São Paulo: Editora 34

Punie Y, Zinnbauer D, Cabrera M (2006) A Review of the Impact of ICT on Learning. Working paper prepared for DG EAC. http://ftp.jrc.es/EURdoc/JRC47246.TN.pdf

Redecker C, Ala-Mutka K, Baciagalupo M, Ferrari A, Punie Y (2009) Learning 2.0: The Impact of Web 2.0 Innovations on Education and Training in Europe. http://ftp.jrc.es/EURdoc/JRC55629.pdf

Robinson K (2011) Out of our minds: learning to be creative. Out of our minds: Learning to be Creative, 2nd edn. Capstone Publishing Ltd., Chichester

Robinson K, Aronica L (2009) The element: how finding your passion changes everything. Viking Penguin, New York

Runco MA, Jaeger GJ (2012) The standard definition of creativity. Creativity Res J. http://doi.org/10.1080/10400419.2012.650092

Shirky C (2014) Why Clay Shirky Banned Laptops, Tablets and Phones from His Classroom. http://www.pbs.org/mediashift/2014/09/why-clay-shirky-banned-laptops-tablets-and-phones-from-his-classroom/

Siemens G (2004) Connectivism: A Learning Theory for the Digital Age. http://devrijeruimte.org/content/artikelen/Connectivism.pdf

Silver EA (1997) Fostering creativity through instruction rich in mathematical problem solving and problem posing. ZDM – Int J Mathe Educ 29(3):75–80. doi:10.1007/s11858-997-0003-x

Speller KG, Schumacher GM (1975) Age and set in creative test performance. Psychol Rep 36(2):447–450. doi:10.2466/pr0.1975.36.2.447

Stein MK, Engle RA, Smith MS, Hughes EK (2008) Orchestrating productive mathematical discussions: five practices for helping teachers move beyond show and tell. Mathe Thinking Learn 10(4):313–340. doi:10.1080/10986060802229675

Tarafdar M, Gupta A, Turel O (2013) The dark side of information technology use. Inf Syst J 23(3):269–275. doi:10.1111/isj.12015

Meaningful Learning in U-Learning Environments: An Experience in Vocational Education

Josilene Almeida Brito[1,2(✉)], Luma da Rocha Seixas[2(✉)],
Ivanildo José de Melo Filho[2,3(✉)], Alex Sandro Gomes[2(✉)], and
Bruno de Souza Monteiro[4(✉)]

[1] IF-SERTÃO – Federal Institute of Sertão Pernambucano, Petrolina, Brazil
josilene.brito@ifsertao-pe.edu.br
[2] Informatics Center, UFPE – Federal University of Pernambuco, Recife, Brazil
{lrs3,asg}@cin.ufpe.br
[3] IFPE – Federal Institute of Pernambuco, Belo Jardim Campus, Belo Jardim, Brazil
ivanildo.melo@belojardim.ifpe.edu.br
[4] UFERSA – Federal Rural University of the Semiarid Region, Mossoró, Brazil
brunomonteiro@ufersa.edu.br

Abstract. Mobile technologies allow new learning experiences anytime, anywhere. In this sense, this study presents a learning experience in urban context supported by a U-Learning environment. Using the Youubi ubiquitous learning platform, the experiment was carried out in a vocational course at a public school in Brazil. Challenging learning situations concerning the urbanization theme of the cities were proposed. The approach adopted was based on the qualitative analysis. The theory of meaningful learning was used as a theoretical framework to assess knowledge construction. The results indicate that the Youubi learning environment fostered the discussions on the content in a dynamic way when outside the classroom, allowing collaboration and knowledge sharing among those involved, mainly strengthening the existing meanings and the perception of problems in the daily life related to the proposed content.

Keywords: Ubiquitous learning · Meaningful learning · Learning strategies · Urban context · Youubi

1 Introduction

The need to improve educational practices that allow learners to place learning in real-world scenarios have been identified by educators for decades (Lave and Wenger 1991; Hung et al. 2013). Many researchers seek to develop learning environments that combine digital resources with world's real-world elements to provide students a real-world experiences with sufficient learning support. For Wu et al. (2013) and Hung et al. (2014) Ubiquitous learning is an approach that allows students to learn from the real world with the support of the learning system using mobile, wireless communication and detection technologies (Hwang et al. 2008).

Ó. Mealha et al. (eds.), *Citizen, Territory and Technologies: Smart Learning Contexts and Practices*,
Smart Innovation, Systems and Technologies 80, DOI 10.1007/978-3-319-61322-2_4

Several studies have demonstrated the benefits of ubiquitous learning in terms of helping students cope with problems as well as the acquisition of knowledge in the real world (Chu et al. 2010; Rogers et al. 2005). For example, Ogata and Yano (2004) developed a ubiquitous context-aware learning system with GPS to guide students to practice in the real world. Hwang and Chang (2011) and Hwang and Tsai (2011) developed a ubiquitous learning environment based on conceptual mapping for conducting field activities in a butterfly garden, while Hwang et al. (2012) has developed a learning system with RFID to guide students to scientific apparatus operate in a science park by assigning several "operational" assess their tasks and operational results.

The authors Hung et al. 2014; Wu et al. 2013 reinforce the importance of promoting learning strategies with personalized learning support in ubiquitous context-aware learning activities. Regarding the sensitivity to the context inherent in the concept of ubiquitous computing, it is emphasized that it is possible to improve the ability to perceive the learners' situation, for example their interests and difficulties, in order to provide them with adaptive assistance in learning activities in everyday situations.

However, we need to question how students appropriate these new technologies and how they impact their study practices. According to Saccol et al. (2011), it is not enough to have access to new and advanced technologies, it is necessary above all to know how to use them to provide learning and development. Consequently, inappropriate u-learning applications can also lead to ineffective and inefficient learning practices. Thus, it is necessary to evaluate how these tools and strategies can be introduced in the teaching-learning activity to achieve the goals of meaningful learning Ausubel (1963).

The adoption of teaching practices based on the meaningful Learning Theory Ausubel (1963) has the purpose of designing learning situations that aim to relate previous knowledge, presented in the classroom, as a meaningful content in the real environment. Thus, for students to learn meaningfully, they must be intentionally engaged in combining prior knowledge with acquired new knowledge Cadorin et al. (2013). Given these possibilities and challenges, this paper describes a ubiquitous approach to meaningful learning based on individual and group practices supported by a ubiquitous learning platform called Youubi. In this direction, an experiment was carried out to investigate the following questions: (1) how ubiquitous learning situations in the form of challenges promote meaningful learning? (2) how students evaluate the Youubi learning environment in the proposed situations?

1.1 Youubi: Ubiquitous Learning Environment

According to Brito et al. (2015) and De Sousa Monteiro et al. (2016) the Youubi environment consists of a client-server architecture. The server provides coordination and communication services through a web service to client applications, and supports social networking, gamification, and user context-based recommendations requirements. To understand the possible scenarios of use of Youubi, it is necessary to analyze its elementary entities: "Person", which represents each user, and "Post", "Event", "Challenge", "Place" and "Group" entities that can be created and manipulated by users. In addition, these entities have geolocation attributes and can be represented by QR (quick response) code, which allows them to associate content to real places.

From this architecture, a mobile phone application client was developed for Android devices, smartphones and tablets. Using this application, teachers and students can create ubiquitous learning situations. For example, they can create and interact with content created and spread throughout the urban space. In addition, students may notice other users nearby, which reduces the feeling of loneliness, both in the virtual environment and in the real environment.

2   Qualitative Method

In order to evaluate how the proposed practices using Youubi promote meaningful learning, an experimental approach has been planned and applied using qualitative analysis techniques. The data collection was based on two sources: (1) interactions and contents created by the participants, obtained through queries to the database of the Youubi server; and (2) questionnaires and interviews that evaluated the student's perception of prior knowledge and after using Youubi.

The context of the study was the urban area of the city of Petrolina, northeastern Brazil. The institution involved was a Federal Public School of vocational education. The educational context was a professional technical courses, and counted with the participation of eighteen (18) students of the technical course Computer Science and twenty (20) students of the Chemistry technical course. The average age of students was nineteen years. The activities were conducted by a Geography teacher in both classes independently, although following the same didactic learning strategies.

2.1   Procedures and Proposed Scenario

The methodological approach consisted of four phases: (1) Initially, was realized a planning of didactic learning activities was carried out, whose theme was Urbanization focusing on "Hydrography and Biome"; (2) then, a pre-test was applied to identify the previous knowledge of the learners; (3) later, the learners started to use the Youubi environment with challenges based on the content discussed in the classroom, (4) finally, a post-test and a questionnaire were applied to collect the students' perceptions about the learning strategy used.

About proposed scenarios, initially, the teacher created challenges (multiple choice questions) in order to present in the platform for the learners and encourage them to participate. The learners were encouraged to resolve them, search for content, collaborate with colleagues in the urban environment, access the QR code, write comments and interact with colleagues through chats.

Then, students were invited to create and share their own content using the Youubi application among the proposed activities. They had to carry out some missions such as taking pictures to illustrate urban problems and publishing them, creating their own posts and questions to challenge colleagues, all related to the topics discussed in their formal learning experiences.

3   Results

In order to identify the students' prior knowledge about the subject, a pre-test diagnostic activity was applied. Considering the answers of 38 apprentices, 22 of them did not present any previous knowledge on the theme "urban space in the contemporary world". Another 9 apprentices attempted to construct knowledge but were misclassified, and finally only 7 apprentices partially built up knowledge. A low level of prior knowledge on the subject in matter was found in most learners.

With respect to activities with the Youubi u-learning environment, these have been designed as opportunities for students to observe and collect information about the urban context, to construct new meanings and to share the discoveries with their colleagues. It is important to note that they used the Youubi environment at home, around the city, and also at school with the guidance of the teacher. In the school were spread QR codes

a) Access to Youubi Content with QR Code.

b) Comments on the Proposed Activities.

c) Challenges Created by Students.

d) Posts Created by Students.

e) Map with the Location of Contents Created in the City.

f) Ranking of the Most Participative Users.

Fig. 1.   Example of learning strategies with Youubi environment.

pointing to contents and challenges created by the teacher (Fig. 1a). These were located in places related to the content of the discipline, with the aim of establishing "learning routes". This didactical strategy allowed to broaden the discussion among students and bring content presented in the classroom to the urban context.

The teacher also created challenges for the learners, with the aim to stimulate interaction and collaboration between them. Thus they were free to discuss and comment teachers' postings and challenges (Fig. 1b). Continuing the experiment, the teacher asked the students themselves to create their challenges (Fig. 1c) and contents (Fig. 1d). The purpose of this activity was to encourage students' autonomy, creativity and critical sense regarding urban problems. All created contents could be consulted at any time on the map (Fig. 1e). Gamification mechanics were also adopted (ranking and medals) to engage students' participation and to foster their development (Fig. 1f).

All the apprentices' interactions were recorded and logged on the Youubi server. The actions of creating, commenting and responding to quizzes were considered the more important actions to understand because they involve a greater cognitive load of apprentices to construct meanings. Table 1 summarizes the most popular actions.

Table 1. Distribution of the main interactions identified on Youubi log.

Private message	1798	Add friend	286	Enter group	81
Reply challenge	1191	Show map	175	Comment challenge	71
Post comment	546	Group messages	123	Create challenge	45
Add post	377	Create post	110	Add place	22

Based on the previous table, you note that learners' interacting with each other through Youubi 1798 times using the chat application. To challenge a quiz also stands out as the one of the most used kind of interaction (n = 1191). This may be related to students' interest in interacting with elements based on game dynamics. It is also noticed that students often used the comment feature to express their opinions about elements already created, whether these simple postings (n = 546), or challenges (n = 71). The interaction with the map (n = 175) to visualize nearby elements appears as a minor common actions. This limitation was maybe occurred caused by the distance between this element and the other entities in the user interfaces of the Youubi client application.

3.1 Meaningful Learning of Ubiquitous Learning Strategy

At the end of the activities carried out with Youubi, a questionnaire using the Likert scale was used to evaluate how meaningful the ubiquitous learning strategy was. The instrument was composed by questions categorized into the four dimensions of meaningful learning: (Q1 and Q2) "motivations to learn", (Q3 and Q4) "interest in learning" (Q5 and Q7), "building shared meanings" and (Q6, Q8 and Q9) "relevance of what they are learning."

This instrument aimed to explore the students' perception about the proposed activities and to find relationships between the ubiquitous learning strategies and the four dimensions of meaningful learning. A summary of the results can be found in Table 2.

Based on the theory of meaningful learning, learners can construct meanings since they are involved in learning activities that generate interest and motivate them to learn. These activities also allow the construction of shared meanings, presenting content relevant to what they learn.

Through the analysis of the answers, we noticed that 97% of the learners answered that they learned with the proposed activity in the environment. About 89% of them agreed that the activities allowed then to share experiences with colleagues. It is interesting to note that 77% answered that they were able to associate the new ideas with their previous experiences and 63% said that authentic materials helped to learn.

Table 2. Summary of students' perceptions regarding aspects of meaningful learning during the experiment with Youubi.

Questions	Average concordance
(Q1) "Activities carried out in the environment helped to learn"	0,97
(Q2) "I was motivated to create activities in the learning environment"	0,74
(Q3) "I felt curious about discussing learning content with real facts"	0,83
(Q4) "I was interested in discussing the learning challenges created and shared by my colleagues"	0,80
(Q5) "I was able to associate the new ideas discussed in the learning activities, relating them to my previous experiences"	0,77
(Q6) "The activity allowed sharing experiences with other people"	0,89
(Q7) "The activity allowed learning with authentic materials related to the real environment"	0,63
(Q8) "The activity performed made me feel less alone when I learned"	0,72
(Q9) "I was able to track learning progress in the proposed activities"	0,69

So, it is possible to conclude that the proposed activities were well accepted by the group participating in the experiment, allowed discussions on urbanization in Petrolina city, contextualized in the place where the student was interactively among the group.

3.1.1   Pre-test and Post-test Analysis

The adoption of the theory of meaningful learning has the objective of designing learning situations that could incorporate connections between the knowledge presented in the classroom and the prior knowledge about the concepts associated with the real urban environment. Analyzing the evolution in the construction of new meanings to achieve meaningful learning we see in Fig. 2 a comparison between previous knowledge on the subject of urbanization with the post-test.

It is noticed that there was a significant increase of apprentices whom 'completely constructed' meanings (7), others 'partially constructed' (18), struggled to respond (13) after the ubiquitous learning didactic activities in the Youubi environment. This conclusion was possible from the analyzes of the experience interpretation of new information and relating it to what you already know.

The theory of meaningful learning also theorizes that in order to assess the learning level of learners, it is necessary to identify the previous knowledge about the new content. The graph lists the evolution in the process of apprehension of the involved learners. Thus, in context-sensitive learning, students learn through the experiences of interpreting new information and relating it to what they already know. As an evidence, students interact with information in the nearby of the classroom, relating the subject studied in the classroom and the situated content.

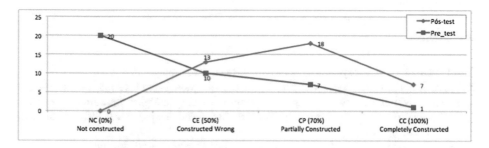

Fig. 2. Comparison of pre-test and post-test results on urbanization.

4   Conclusion

The paper presented the main results of an experiment on teaching urbanization in a technical training course, designed using the Youubi learning environment to support meaningful learning, using geolocation features and QR code tags. In this study, we assessed effectiveness of this ubiquitous didactic learning strategies to promote meaningful learning.

The results show that the didactic strategies can improve learning. Given the identified data regarding the level of satisfaction of use, we conclude that the environment presented moments of satisfaction and moments of disagreement. Although in all the questions there was a positive index of acceptance of the same, it was highlighted the satisfaction acquired in using the ubiquitous technologies available in the environment even presenting some moments of disagreement. Also highlights the involvement and knowledge of school teachers in the planning and development of activities based in the real world, specified in learning environments; Therefore, it is necessary to promote a greater adequacy of mobile learning environments so that teachers can develop activities in an easy, more dynamic way involving real-world content.

The experimental results showed that, with the help of ubiquitous learning strategies, the learners presented a very significant performance in carrying out the activities. The learners' feedback about their level of involvement in the conceptual dimensions of meaningful learning: "motivation to learn", "intention/Interest in learning", "construction of shared meanings", and "relevance to the applicability of authentic materials".

Majority of learners considered positive the learning experience. So, we conclude that the ubiquitous didactic learning strategies in the form of learning challenges seems to help students to organize and refine their field observations and knowledge construction.

Although there are still limitations in the graphical interface of the Youubi learning environment, pointed out in disagreement on the ease of use of the environment, the ubiquitous technologies used as resources in the learning challenges were considered positive.

In the near future, we intend to further improve the interface to improve communicability and usability with new features that allow greater analysis and feedback of learners' performance as well as engaging them in more interactive and meaningful learning situations.

References

Ausubel DP (1963) The psychology of meaningful verbal learning

Brito JA, Amorim R, De Sousa Monteiro B, Gomes AS, Melo Filho IJ (2015) Effectiveness of practices with sensors in engaging in meaningful learning in higher education: extending a framework of ubiquitous learning. In: Frontiers in Education conference (FIE), 2015. 32614 2015. IEEE, pp 1–4

Cadorin L, Bagnasco M, Rocco G, Sasso L (2013) Meaningful learning in healthcare professionals: integrative review and concept analysis. In: The inaugural European conference on education 2013 official conference proceedings. Brighton, UK

Chu HC, Hwang GJ, Tsai CC (2010) A knowledge engineering approach to developing mindtools for context-aware ubiquitous learning. Comput Educ 54(1):289–297

De Sousa Monteiro B, Gomes AS, Neto FMM (2016) Youubi: open software for ubiquitous learning. Comput Hum Behav 55:1145–1164

Hung IC, Yang XJ, Fang WC, Hwang GJ, Chen NS (2014) A context-aware video prompt approach to improving students' in-field reflection levels. Comput Educ 70:80–91

Hung PH, Hwang GJ, Lin YF, Wu TH, Su IH (2013) Seamless connection between learning and assessment-applying progressive learning tasks in mobile ecology inquiry. Educ Technol Soc 16(1):194–205

Hwang GJ, Chang HF (2011) A formative assessment-based mobile learning approach to improving the learning attitudes and achievements of students. Comput Educ 56(4):1023–1031

Hwang GJ, Tsai CC, Yang SJ (2008) Criteria, strategies and research issues of context-aware ubiquitous learning. Educ Technol Soc 11(2):81–91

Hwang GJ, Tsai CC (2011) Research trends in mobile and ubiquitous learning: a review of publications in selected journals from 2001 to 2010. Br J Educ Technol 42(4):E65–E70

Hwang GJ, Tsai CC, Chu HC, Kinshuk K, Chen CY (2012) A context-aware ubiquitous learning approach to conducting scientific inquiry activities in a science park. Australas J Educ Technol 28(5)

Lave J, Wenger E (1991) Situated learning: legitimate peripheral participation. Cambridge University Press, Cambridge

Ogata H, Yano Y (2004) Context-aware support for computer-supported ubiquitous learning. In: The 2nd IEEE international workshop on wireless and mobile technologies in education, 2004. Proceedings. IEEE, pp 27–34

Rogers Y, Price S, Randell C, Fraser DS, Weal M, Fitzpatrick G (2005) Ubi-learning integrates indoor and outdoor experiences. Commun ACM 48(1):55–59

Saccol A, Schlemmer E, Barbosa J, Hahn R (2011) M-learning e u-learning: novas perspectivas da aprendizagem móvel e ubíqua. Pearson Prentice Hall, São Paulo

Wu HK, Lee SWY, Chang HY, Liang JC (2013) Current status, opportunities and challenges of augmented reality in education. Comput Educ 62:41–49

The Classroom Physical Space as a Learning Ecosystem - Bridging Approaches: Results from a Web Survey

Lara Sardinha[1(✉)], Ana Margarida Pisco Almeida[1(✉)],
and Maria Potes Barbas[2]

[1] CIC.Digital/DigiMedia, University of Aveiro, Aveiro, Portugal
{larasardinha, marga}@ua.pt
[2] Santarém Higher School of Education, Santarém, Portugal
mariapbarbas@gmail.com

Abstract. The classroom physical space enfolds several dimensions such as the social, cultural, architectural and technological. The current scenario of digitally equipped classrooms in which new pedagogical approaches based on collaborative learning, project-based learning and personalized learning are being used, call for the need to rethink the classroom physical space. Despite of the existence of some new classroom physical spaces aiming to answer this new reality, like the Future Classroom Lab, we argue that there might be lacking an innovative interior design strategy encompassing these aspects and fulfilling all the classroom physical space dimensions. Thus, this paper aims to present the perspective the authors have concerning the classroom physical space as a learning ecosystem and to start building the bridges between different approaches to space and relating them to the classroom physical space, in order to create an innovative interior design strategy that will improve the use of classroom physical space in its different dimensions. We also present the first results of an European web survey applied to the European Schoolnet Future Classroom network members that aimed at understanding how their spaces were thought and how they are being perceived; a brief discussion of the results, which, overall, are positive, is also presented. The paper ends with some references to the future work.

Keywords: Classroom physical space · Smart learning ecosystems · Classroom orchestration · Enabling spaces · Human-building interaction · Smart classroom · Spatial semiotics · Spatial pedagogy

1 Introduction

This paper is part of an ongoing research that aims to investigate the role of innovative interior design strategies in creating new classroom spaces. Acknowledging that the Classroom Physical Space (CPhS) interacts and depends directly on several different dimensions among which the social, cultural and digital, a new space (Sardinha et al. 2017) is going to be designed aiming to promote the inclusion of specific populations, namely the youngsters that are Not in Education, Employment or Training (NEET) and Refugees.

© Springer International Publishing AG 2018
Ó. Mealha et al. (eds.), *Citizen, Territory and Technologies: Smart Learning Contexts and Practices*,
Smart Innovation, Systems and Technologies 80, DOI 10.1007/978-3-319-61322-2_5

Several approaches to the space are introduced: classroom orchestration, the enabling spaces, Human-Building Interaction (HBI), the smart learning ecosystem and smart classrooms, in order to start creating bridges among them.

Some of the data already collected through an European web survey regarding the use of the Future Classroom Learning Labs (FCLL) is also presented.

2   The Classroom Physical Space

The research on the classroom physical space involves the development of a multi-disciplinary approach that must consider different dimensions and contributes from several domains as the classroom orchestration (Dillenbourg and Fischer 2007; Dillenbourg and Jermann 2007), the enabling spaces approach (Peschl and Fundneider 2012), the HBI (Alavi et al. 2016a, b) and the spatial semiotics and spatial pedagogy (Lim et al. 2012). In our perspective, all these have in common the high relevance given to the social dimension of the CPhS that, together with the technological one alongside with the spatial semiotics and the spatial pedagogy plays a relevant role in the creation of a smart learning ecosystem.

When considering the specific target population of the project that frames this paper (NEET and Refugees), it seems relevant to better understand how can these dimensions shape a new context and help to create smart spaces that might potentially enhance a more inclusive and better CPhS, i.e., "*a context where the human capital (and more in general each individual) owns not only a high level of skills, but is also strongly motivated by continuous and adequate challenges, while its primary needs are reasonably satisfied*" (Giovannella 2014a, b).

2.1   A Contextualization of the Classroom Physical Space History

According to Park and Choi (Park and Choi 2014) the classroom physical space has been connected to the educational approaches through time. In ancient Greece there was a rhetorical/dialogical system and there were neither a specific space for the classes to happen, nor a rigid setting for the teacher and the students to be. These latter would place themselves around the teacher in no particular order.

When a more formal education appeared with the medieval Universities, a more rigid layout set place and evolved to a very strict layout with the spreading of Universities. During this time, the educational system was teacher-centred, and the classroom layout reflected this centrism, occupying the teacher a featured place in the classroom. With the expansion of Universities and schools, the medieval layout remained, however adapted to a bigger space (Fig. 1).

In the last century the pedagogical approaches started to change, although the classroom space and layout, in general, did not reflected these changes. Towards the end of the XX century and, in particular, in the beginning of the present one, the CPhS started to be reconfigured. Not only is this change of paradigm due to the technological penetration in the classrooms, but also to new pedagogical approaches that came with it. The SCALE-UP (Burke 2015) space and the Future classroom Lab (FCL) (European Schoolnet 2016) are good examples which translate this educational shift.

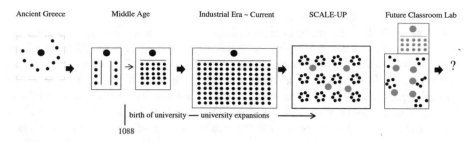

Fig. 1. Adaptation of the fig. "Historical changes in classroom design" by Park and Choi (2014)

2.2 The Space Approaches to Create Bridges

In our perspective, and alongside with some other investigators (Dillenbourg et al. 2011; Peschl et al. 2014), the classroom is a complex system that combines different dimensions as the social, cultural, architectural and digital, among others. Even if there is a mental image of "four delimiting walls" connected to the classroom space, we approach this latter as going further and outside the walls. Thus, the way the inside walls is thought and designed should broaden it across its physical boundaries. In order to comply this, digital technologies can play a very important role in creating new scenarios that can better enable the learning processes, through a technological enhanced environment (Giovannella 2014a).

2.2.1 Spatial Semiotics and Spatial Pedagogy

Space is a way of communication and in order to understand the way it flows through space distance (Hall 1959) and its dynamics (Stenglin 2009), it is important to master its language. The built edification can be defined through: (a) an Euclidean perspective as the architectural space (Peschl and Fundneider 2012) or the built space (Stenglin 2009) relying, in part, in static and dynamic resources (Stenglin 2009); (b) the "constructed" space between walls which cannot be detached from a social dimension brought up by the interaction and interpersonal relations between its users (Stenglin 2009; Perolini 2011). This social dimension, together with a cultural one, reflects on how space is organized (Hall 1959) as each culture experiences it in its own way (Hall 1990).

In which concerns CPhS, and more specifically to the classroom spaces conceived meanings, a way of experiencing space is often times present in the interactions between teachers and students and space itself, or spatial semiotics (Lim et al. 2012). This latter is perceived through the way they move athwart space and its signification and their paths (Lim et al. 2012). The different paths and their quality constructed through movement across the space may measure fluidity.

The study of the movements and its positions and directions, allow the arising of patterns enabling the analysis of the dynamic of the physical space. The positioning and directionality of movements in a classroom usually are not random, having a meaning, as well as face expressions, gestures and the voice intensity. These encompass a semiotics dimension, which alongside with the language and pedagogical ones, among others, define spatial pedagogy (Lim et al. 2012).

2.2.2 The Classroom Orchestration

Classroom orchestration, and in particular, Dillenbourg's perspective of it, ensue from the ability to manage a technological enhanced environment (like the Computer Supported Collaborative Learning environment - CSCL), not only through the core of instructional design (kernel) but also through observing learners during their activities and making the necessary adjustments to learners instructions whenever needed (towards personalization) (Dillenbourg et al. 2011). To Dillenbourg, the "rings around the kernel" cannot be neglected, even those that might look out of place as the rings which address logistics. Though, Dillenbourg argues the existence of constraints to both kernel and rings, being the kernel constraints related to: what (the curriculum), what is inside (the contents), and who (how people learn as well as the learners themselves). What regards the rings constraints, or the "designing for orchestration", Dillenbourg considers these to be constraints related to the assessment, time, discipline, energy and space (Dillenbourg et al. 2011). All these constraints both of kernels and rings' have a deep influence on the teacher's work, meaning that the author considers the teacher's role as having the most importance in classroom orchestration and proposes their empowerment, in order to increase student's achievements via problem solving situations and group discussions. This empowerment should not be achieved by simply placing the teacher as the commanding agent, but as the one that scaffolds and enhances students' motivation towards successful achievements, through design factors (leadership, flexibility, control, awareness, etc.).

Classroom orchestration in the referred perspective aims to provide a better learning ecosystem (physical, technological, social, personal, emotional) to students in order to scaffold and enhance their knowledge acquisition.

2.2.3 The Enabling Spaces Approach

Peschl and Fundneider define Enabling Spaces as multidimensional spaces (architectural, social, emotional and technological dimension spaces, among others), which enable, facilitate and support the knowledge creation and innovation processes. For the authors, the optimization of new knowledge creation is empowered by the multidimensional spaces, each one corresponding to a different dimension, which must be "orchestrated in an integrated manner" (Peschl and Fundneider 2012) as well as in an interdisciplinary way overcoming the possible constraints and conditions (Peschl and Fundneider 2012).

When applied in the educational context, the Enabling Spaces approach also encompasses the pedagogical choices and the didactical environment, as well as the teachers' personality (beliefs and thoughts) alongside with the different spaces/dimensions mentioned above (Peschl et al. 2014).

Therefore, in the Enabling Spaces approach "the integration and orchestration of different spaces/dimensions (…) is one of the most challenging problems, yet powerful features" (Peschl and Fundneider 2012). In order to overcome this challenges, Peschl and Fundneider stress the importance of supporting and leading the Enabling Space interdisciplinary through a well-founded design process (Peschl and Fundneider 2012).

2.2.4   Human-Building Interaction

HBI brings forward the relation between Human-Computer Interaction and buildings. As these latter are becoming more and more technological based, like in the Smart Homes, Alavi et al. (2016b) argue that buildings ought to be developed and designed with a dialogical relation between its users (either in the social and individual levels) and their "digital and physical interactive daily experiences" (Alavi et al. 2016b).

HBI approaches buildings through Hillier's perspective (Alavi et al. 2016a, b) in which besides the physical and spatial form these also have a social-cultural function (Hillier 2007).

According to the HBI authors, "Designing HBI (...) consists of providing interactive opportunities for the occupants to shape the physical, spatial, and social impacts of their built environment" (Alavi et al. 2016b).

2.2.5   Smart Learning Ecosystems and the Smart Classroom

Smart learning ecosystems encompasses not only the students, teachers and school staff as "individual actors of the learning process" (Galego et al. 2016) but also the stakeholders, surrounding community, family, "services, social life, challenges, skills" (Galego et al. 2016) inherent to the learning environment. Smart learning ecosystems, apart from the smart technology, devices, applications and its infrastructures, relies also in "help[ing] towards achieving a people centred smartness, through streamlining mundane organisational tasks, and enhancing the skills of all actors involved in learning processes" (ASLERD 2016).

Bautista and Borges state that the concept of smart classroom arises from the intersection between "classroom's architectural design and its ergonomy", smart technology and pedagogical approaches "as collaborative learning, project-based learning, (...) students' autonomy, educational co-responsibility, etc." (Bautista and Borges 2013) relying also on the actors' learning processes.

2.2.6   Bridging the Space Approaches

Physical spaces when detached from their social and cultural dimensions risk to lose their meaning. Thus, when approaching the CPhS, we intend to create bridges between the referred space dimensions and to develop an innovative interior design strategy to the CPhS. In this process, we will give relevance to the social and cultural meanings of the physical space, as well as to the interaction opportunities and the state of flow of all agents involved. To study these interaction opportunities is central and we must consider three main dynamics: between users and technology, between users and space and among users themselves.

3   The Web Survey

We applied a web survey aiming to understand how the FCLL from the European Schoolnet FCL network members were thought and conceived and how these are being used. It targeted the Decision Makers (DM), the Decision Makers which are also Teachers (DMT), Teachers (T) and Students (S).

Section 1 was presented to characterize the participants and it was common to all the groups, as well as section 6. This intended to gather a list of the technological solutions in use in the FCLL, as well as to understand how the FCLL layout is being displaced.

Section 2 and part of section 3, targeting DM and DMT, aim to understand what reasons/factors led to the decision making of implementing a FCLL in the school and how these spaces were thought/conceived.

The remaining sections 3 and 4, this latter targeting teachers, is almost identical to the students' section (section 5). These aim to understand how the FCLL are being used by teachers and students and their perception of it through a 5-point Likert scale (strongly disagree to strongly agree). The questions were categorized in physical space, space communication, emotional space, teaching/learning space, social space and technological space.

3.1   The Participants

The online web survey dissemination was done by email and Facebook. An email was sent to all the contacts available on the FCL website in November 2016 and to the European Schoolnet. There were 26 FCLL from 12 countries of which: Portugal (9); Belgium (4); Germany, Israel and Norway (2); Croatia, Cyprus, Czech Republic, France, Italy, Slovakia and United Kingdom (1). They were asked to spread the web survey to all the DM, DMT, T and S using the FCLL. The European Schoolnet posted the link to the web survey on their Facebook page.

107 complete questionnaires were collected, from which: 3 DM (3%), 10 DMT (9%), 11 T (10%) and 83 S (78%). To what concerns the gender, despite 82% being male, if we consider the DM, DMT and T alone, then we have 67% being female. The age mean is 23 years old with a standard deviation (SD) of 12; however the age mean concerning the DM, DMT and T is 44 years old (SD 9,16), being the oldest 66 years old and the youngest 32 years old. The students' age mean is 17 years old (SD 1,59), being the oldest 25 years old and the youngest 15 years old.

Most of the participants, 94%, are from Portugal (101) including all the students, being the other 6 participants from Belgium (1), France (1), Israel (1), Italy (2) and Norway (1). In what concerns the type of school where the FCLL are located is worth to mention that 82% (89/107) of the respondents, of which 82 students, are from the same school – a Portuguese VET School[1]. The others are: Elementary School (1), Middle School (4), High School (4), University (4), Norwegian Education Government (1), Showcase (1), ICT Centre (1), Teacher Training Centre (1) and Cluster of Schools (1).

---

[1] The FCLL from this school opened in September 2016 and the person responsible has shown quite some enthusiasm for participating in this study.

3.2 Results and Discussion

The results hereinafter presented concern to the relative frequency of the quantitative data collected and are organized according to two main items: factors leading to a decision of implementing a FCLL and the use of the FCLL and users' perception of it. Other obtained results are not detailed in this paper.

3.2.1 Factors Leading to a Decision of Implementing a FCLL

The analysis of the factors that led to the decision-making concerning the FCLL implementation was made considering the answers given by the DM and the DMT (13/107) from 12 different FCLL.

According to our data, the principal factor that led the decision makers to implement the FCLL in their school was the *Future Classroom pedagogical approaches* (6) followed by the reason of the schools' *Students with learning difficulties* (3) and the *School philosophy* (2). Two other factors have been pointed out: the *pedagogical needs* (1) and the fact of *Taken part of ITEC Project* (1).

Despite of what 11 decision makers have said that the FCLL of their school is inspired by the Brussels FCL layout[2], only 9 of these are based on the Brussels FCL layout despite having quite some differences. Nevertheless, the identified main reasons for their FCLL being different from the one in Brussels were: the budget (7), the chosen physical space not being the most suitable (6), the School culture (5) and the specificity of the School's students (5). 6 of the 13 decision makers also stated that their FCLL has an area that differs from the Brussels FCL like a playing/gaming area.

3.2.2 The Use of the FCLL and Users' Perception of It

Some questions of the web survey regard the physical and communicative space. The participants (104/107) have an overall positive perception of the initial use of the FCLL; 77% consider the FCLL space to be intuitive (Table 1) and 81% think that was easy to identify the different areas (Table 2).

Table 1. Perceptions towards the FCLL space being intuitive

	Negative perception	Neutral perception	Positive perception
DMT	20.0%	30.0%	50.0%
T	9.1%	18.2%	72.7%
DMT&T	14.3%	23.8%	61.9%
S	4.8%	14.5%	80.7%
DMT&T&S	6.7%	16.3%	76.9%

---

[2] The Brussels FCL layout comprises six learning zones in two different spaces: (1) one space based on the traditional classroom furniture setting, the interact learning zone, and (2) the remain-ing five learning zones (create, present, investigate, exchange and develop) are organized through an open space equipped with different type of technology.

Table 2. Perceptions towards the easiness in identifying the different areas in the FCLL

	Negative perception	Neutral perception	Positive perception
DMT	20.0%	10.0%	70.0%
T	0.0%	18.2%	81.8%
DMT&T	9.5%	14.3%	76.2%
S	4.8%	13.3%	81.9%
DMT&T&S	5.8%	13.5%	80.8%

However, even though 83% think that it was easy to adapt to (Table 3) is interesting to notice that 40% of the DMT had a negative perception of it, the exactly same amount for the positive perception stated by them.

Table 3. Perceptions towards the easiness in adapting to use the FCLL space

	Negative perception	Neutral perception	Positive perception
DMT	40.0%	20.0%	40.0%
T	0.0%	18.2%	81.8%
DMT&T	19.0%	19.0%	61.9%
S	4.8%	7.2%	88.0%
DMT&T&S	7.7%	9.6%	82.7%

Nevertheless, only 56% say that there was no need to have an explanation on how to use the FCLL space against 21% of a negative perception (Table 4). It is also interesting to notice that when separating S (83/107) from DMT&T (21/107) the values differ in more than 20% for the need to have an explanation on how to use the space – 38% DMT&T against 17% of the S.

Table 4. Perceptions towards the need to have an explanation on how to use the FCLL space

	Negative perception	Neutral perception	Positive perception
DMT	40.0%	30.0%	30.0%
T	36.4%	9.1%	54.5%
DMT&T	38.1%	19.0%	42.9%
S	16.9%	24.1%	59.0%
DMT&T&S	21.2%	23.1%	55.8%

When focusing the questions on the FCLL layout organization towards teaching and learning, the results are in general more alike and positive: when asked about if *in the FCLL it is easy to pass from an activity area to another without disturbing the students/classmates,* we have for positive perception 76% for DMT&T and 67% for S these latter have a slight more positive opinion towards the *spatial FCLL organization allowing them to understand which kind of activity they are about to start* – 73%, despite DMT&T remaining in the same percentage of 76% when asked if *the spatial*

FCLL organization allows them to explain which kind of activity they are about to start. Yet, when questioned if *the spatial FCLL organization is suitable for different kind of activities*, the opinion differ again – 75% of the S had a positive perception of this statement against 95% of the DMT&T.

When the question refers to the *facility of moving the FCLL furniture according to the different activities*, the positive perception presents a decrease to 66% of the S and 81% of the DMT&T. We find a slight difference of the positive perception in which regards the *ease to transform the FCLL layout (furniture displacement)* 68% and 76% (S and DMT&T, namely) and a slight increase of the positive perception to 74% (S) and 86% (DMT&T) when asked about *the activities being enabled by the existing furniture in the FCLL.* In what concerns *the furniture used in the FCLL enabling the teaching improvement and learning improvement* the S and DMT&T's positive perception were akin to the previous one, 66% and 71%, namely, regarding the learning improvement and 72% (S) and 67% (DMT&T) for the other ones. In what regards *the FCLL existing furniture being the most suitable for teaching* the DMT&T positive perception stays at 57% and the S' positive perception in 58%. When questioned about *the FCLL furniture being the most suitable for learning*, the positive perceptions are set in 57% (DMT&T) and 66% (S). Still, when enquired if *the activities were enabled by the FCLL existing layout*, 86% of the DMT&T and 70% of the S had a positive perception.

Some results differ to what concerns *the FCLL layout enabling* (1a) *the teaching improvement* (Table 5) and (1b) the *student improvement* (Table 6) and to what regards *the FCLL layout being the most suitable* (2a) *for teaching* (Table 7) and (2b) *for learning* (Table 8). It is worth to notice the difference of positive perceptions not only between the DMT&T and S but also between (1ab) and (2ab).

Table 5. Perceptions on how the FCLL layout enables the teaching improvement

	Count	Negative perception	Neutral perception	Positive perception
DMT	10	0.0%	20.0%	80.0%
T	11	0.0%	9.1%	90.9%
DMT&T	21	0.0%	14.3%	85.7%
S	83	2.4%	37.3%	60.2%
DMT&T&S	104	1.9%	32.7%	65.4%

Table 6. Perceptions on how the FCLL layout enables the learning improvement

	Count	Negative perception	Neutral perception	Positive perception
DMT	10	0.0%	20.0%	80.0%
T	11	9.1%	18.2%	72.7%
DMT&T	21	4.8%	19.0%	76.2%
S	83	6.0%	27.7%	66.3%
DMT&T&S	104	5.8%	26.0%	68.3%

Table 7. Perceptions on how the FCLL layout is the most suitable for teaching

	Count	Negative perception	Neutral perception	Positive perception
DMT	10	0.0%	40.0%	60.0%
T	11	0.0%	27.3%	72.7%
DMT&T	21	0.0%	33.3%	66.7%
S	83	2.4%	37.3%	60.2%
DMT&T&S	104	1.9%	36.5%	61.5%

Table 8. Perceptions on the FCLL layout is the most suitable for learning

	Count	Negative perception	Neutral perception	Positive perception
DMT	10	0.0%	30.0%	70.0%
T	11	9.1%	45.5%	45.5%
DMT&T	21	4.8%	38.1%	57.1%
S	83	2.4%	30.1%	67.5%
DMT&T&S	104	2.9%	31.7%	65.4%

A better positive perception (S and DMT&T) regarding the same range of questions but instead of the layout or the furniture, they are questioned about the existing technology in the FCLL, still being noticed a difference between the two groups: *enabling the teaching improvement* the positive perceptions are 77% (S) and 86% (DMT&T); *enabling the learning improvement,* 65% (S) and 76% (DMT&T); *the activities being enabled by the existing technology in the FCLL* present a positive perception from the S of 70% and from the DMT&T of 86% and it what regards *the FCLL existing technology being the most suitable for teaching* and *for learning* we have, positive perceptions of 70% (S) and 81% (DMT&T) for the teaching. For the learning the S' positive perception is the same however the DMT&T' positive perception decreases for 71%.

Despite the FCLL being designed to allow different spatial configurations in a regular basis, only 42% of the participants (45/107) of the web survey say that *in their FCLL the layout changes,* and from these, 49% is occasionally and 26% once a week to daily; being usually either the teachers (40%) or the students together with the teachers (40%) changing the layout.

3.2.3   Discussion

In general, and from this initial web survey, we may say that the current scenario regarding the FCLL physical space is positive as their users have a positive perception of it. However, from the results, we may infer that the potential of the FCLL physical space is not at its best. Results showed that the options made by the decision makers took into consideration the CSCL and project-based learning approaches present in the Future Classroom project (Van Assche et al. 2015) as well as a social concern regarding the school's population. Nevertheless, and even if the decision makers have taken into account the families, community, stakeholders, as in a smart learning ecosystem, none of the results support this aspect.

Regarding the physical space, its communication and use, results are irregular, particularly if we consider the S and the DMT&T groups separately. Through a HBI perspective, the dialogical relations as well as the users' built environment shaping through interactive opportunities seem not to be completely adjusted as the perceptions users have, despite being positive, present imparities: 77% of the FCLL users state that the space is intuitive, however 56% said that an explanation how to use the space was required; or, the disparities concerning the furniture and the space layout mentioned above. From these results we also might argue that the "balance" between the different space dimensions, and in particular the architectural and the technological ones, is not the most consistent.

Therefore, we argue that an innovative interior design strategy regarding the CPhS and bridging the different presented approaches is in order.

4 Future Work

Apart from the analysis of the web survey qualitative data, a correlational statistic regarding these latter is being conducted, in order to gather more grounded information to create the scripts for both the interviews and the workshop/focus group. The participants of these interviews will be key-elements connected to the Portuguese FCLL as its objective is to consolidate some of the data already collected, as well to gather more data regarding the classroom physical space. NEET/Refugee population will participate in the workshop/focus group where they will be asked to design classroom spaces followed by a group discussion.

References

Alavi HS, Churchill E, Kirk D et al (2016a) Deconstructing human-building interaction. Interactions 23:60–62. doi:10.1145/2991897

Alavi HS, Lalanne D, Nembrini J et al (2016b) Future of human-building interaction. In: Proceedings of the 2016 CHI conference extended abstracts on human factors in computing systems - CHI EA 2016. ACM Press, New York, pp 3408–3414

ASLERD (2016) TIMISOARA DECLARATION better learning for a better world through people centred smart learning ecosystems

Bautista G, Borges F (2013) Smart classrooms: innovation in formal learning spaces to transform learning experiences. Bull IEEE Tech Committeee Learn Technol 15:18–21. http://lttf.ieee.org/

Burke DD (2015) Scale-Up! classroom design and use can facilitate learning. Law Teach 49:189–205. doi:10.1080/03069400.2015.1014180

Dillenbourg P, Fischer F (2007) Basics of computer-supported collaborative learning. Zeitschrift für Berufs- und Wirtschaftspädagogik 21:111–130

Dillenbourg P, Jermann P (2007) Scripting computer-supported collaborative learning. Springer, Boston

Dillenbourg P, Sharples M, Fischer F et al (2011) Trends in orchestration. Second research and technology scouting report

European Schoolnet (2016) Future classroom lab learning zones. 1–10

Galego D, Giovannella C, Mealha Ó (2016) An investigation of actors' differences in the perception of learning ecosystems' smartness: the Aveiro University case. Interact Des Archit - Proc 1st Smart Learn Ecosyst Reg Dev Conf 31:1–13

Giovannella C (2014a) Smart learning eco-systems: 'fashion' or 'beef'? J e-Learning Knowl Soc 10:15–23

Giovannella C (2014b) Where's the smartness of learning in smart territories? Smart city learn oppor challenges - EC-TEL 2014 work 1–6

Hall ET (1959) The silent language. Double & Day Company, Inc., Garden City

Hall ET (1990) The hidden dimension, reprint. Anchor Books

Hillier B (2007) Space is the machine, electronic. Space Syntax, London

Lim FV, O'Halloran KL, Podlasov A (2012) Spatial pedagogy: mapping meanings in the use of classroom space. Cambridge J Educ 42:235–251. doi:10.1080/0305764X.2012.676629

Park EL, Choi BK (2014) Transformation of classroom spaces: traditional versus active learning classroom in colleges. High Educ 68:749–771. doi:10.1007/s10734-014-9742-0

Perolini PS (2011) Interior spaces and the layers of meaning

Peschl MF, Bottaro G, Hartner-Tiefenthaler M, Katharina R (2014) Learning how to innovate as a socio-epistemological process of co-creation. Towards a constructivist teaching strategy for innovation. Constr Found 9:27

Peschl MF, Fundneider T (2012) Spaces enabling game-changing and sustaining innovations: why space matters for knowledge creation and innovation. J Organ Transform Soc Chang 9:41–61. doi:10.1386/jots.9.1.41_1

Sardinha L, Almeida AMP, Barbas MP (2017) Digital future classroom: the physical space and the inclusion of the NEET/Refugee population - conceptual and theoretical frameworks and methodology. In: INTED 2017 - 11th annual international technology, education and development conference. IATED, Valencia, Spain, pp 2396–2405

Stenglin MK (2009) Space odyssey: towards a social semiotic model of three-dimensional space. Vis Commun 8:35–64. doi:10.1177/1470357208099147

Van Assche F, Anido-Rifón L, Griffiths D et al (eds) (2015) Re-engineering the uptake of ICT in schools. Springer International Publishing, Cham

The Power of the Internet of Things in Education: An Overview of Current Status and Potential

Filipe T. Moreira[⊠], Andreia Magalhães, Fernando Ramos,
and Mário Vairinhos

Department of Communication and Art, CIC.Digital/DigiMedia,
University of Aveiro, Aveiro, Portugal
{filipertmoreira, fernando.ramos, mariov}@ua.pt,
andreiamagalhaes78@gmail.com

Abstract. This paper aims at discussing the potential and analyze main challenges concerning the use of the Internet of Things in education, based on an overview of the literature and the identification of some relevant technologies and projects. The integration of Internet of Things technologies, including open data, in schoolbooks, identified in this context as smartbooks, is also address. Regarding open data, a set of data sources is drawn and possible scenarios for integrating these data into the manuals are discuss. It is concluded that although the Internet of Things presents the potential to facilitate access to real and constantly updated data, there is still a long way to go in this field, namely, in primary and secondary education.

Keywords: Internet of Things · Education · Open data

1 Introduction

In recent years, the Internet of Things (IoT) has been attracting increasing interest, as can be seen in the Internet search indexes or in the last editions of the Gartner's hype cycle for emerging technologies. Regarding scientific investigation, in its in-depth analysis of IoT solutions on the market (Pereira et al. 2015) note that the development of research and investment of IoT solutions, as well as projects in the academic field, is being conduct in five main categories: smart wearable, smart home, smart city, smart environment, and smart enterprise.

However, despite the interest in this technology, it seems to exist some obstacles in its implementation. According to Mukhopadhyay and Suryadevara (2014) these obstacles are related to the still lack of understanding of the application and real usefulness of IoT products or services. Other aspects referred as obstacles are: a lack of commitment and knowledge of higher management; the lack of maturity of IoT industry standards; and, the still high costs of IoT infrastructures.

© Springer International Publishing AG 2018
Ó. Mealha et al. (eds.), *Citizen, Territory and Technologies: Smart Learning Contexts and Practices*, Smart Innovation, Systems and Technologies 80, DOI 10.1007/978-3-319-61322-2_6

2   The Internet of Things

This section presents a definition of IoT and a brief historical overview of its development that will help better contextualize the subject.

2.1   Concept

There is currently no broadly accepted conceptual definition of IoT, however, generally speaking, this term refers to physical objects - or things - that incorporate electronics, software, sensors and network connectivity, enabling the things to collect and share data. This allows these objects to be monitor and controlled remotely, creating new opportunities for a seamless integration between the physical and digital worlds (Ray et al. 2016). The basic idea is that behind this ubiquitous presence of objects/things surrounding humans, these objects/things are able to measure, infer, understand and even modify and act in the environment in which are inserted (Bota et al. 2016).

In a somewhat simplistic way, IoT allows individuals and things to be connect at any time, in any place, and with anyone, ideally using any path/network and any service (Perera et al. 2014).

For O'Brien (2016), IoT is defined as a technology that allows, through sensors, to connect objects with the Internet, so that information from any given environment or activity can be obtained and stored – information which will then provide feedback and allow for better control over those same environments and activities. However, in a disruptive view of the future, the author even affirms that IoT "has been identified as the third wave of the Internet. It has also been identify as the fourth industrial revolution "(O'Brien 2016).

In a more systematic definition of Ray et al. (2016), IoT is an ecosystem that exploits and expands widely existing environments of embedded and connected devices. The scope of the IoT is the computing infrastructure that will allow an ecosystem in which there are more "things" connected to the Internet than the number of people (Ray et al. 2016).

Among the most cited authors when addressing the definition of IoT are Atzori et al. (2010), which present a definition based on the cross-referencing of three paradigms - an "Internet"-oriented paradigm (middleware), a "Things"-oriented paradigm (sensors) and a "Semantics"-oriented paradigm - represented in Fig. 1. This type of definition arises as necessary due to the interdisciplinary nature of the subject; however, the usefulness of the Internet of Things can be trigger only in an application domain where the three paradigms intersect (Gubbi et al. 2013).

With a deep insight of the term, Sundmaeker and Saint-Exupéry (2010) analyze the meaning of the word "things" through an Aristotelian philosophic perspective, concluding that the term is not restricted to material objects. Thus, the authors state that "things" can be define as real/physical or digital/virtual entities that exist, that move in space and time and are capable of being identified. For these authors, IoT is:

> *"a dynamic global network infrastructure with self-configuring capabilities based on standard and interoperable communication protocols where physical and virtual "things" have identities, physical attributes, and virtual personalities and use intelligent interfaces, and are seamlessly integrated into the information network" (p. 43).*

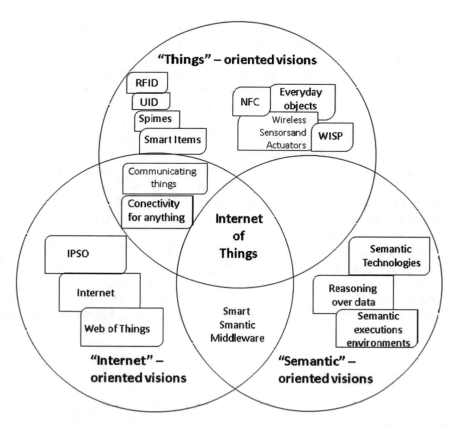

Fig. 1. 'Internet of Things" paradigm as a result of the convergence of different views (based on Atzori et al. 2010)

This interpretation of IoT not as a technology but as a global concept is also advocate in the 2016 report of the IERC - IoT European Research Cluster (Vermesen and Friers 2016).

According to Xia et al. (2012) "IoT refers to the networked interconnection of everyday objects, which are often equipped with ubiquitous intelligence". In the same paper, the authors say that these technologies will increase the ubiquity of the internet because they will integrate all objects into an embedded system, which will give rise to a widely present network of objects communicating with humans or other objects.

2.2   Brief Historical Overview

It is possible to divide the Internet's evolution into four stages: the Web, aimed at connecting and obtaining information in the network; the Social Web, characterized by the concern of the active participation of the user and the collaboration through the social networks; the Semantic Web, with concentrated efforts in the attribution of meaning and context to information; and the *Ubiquitous Web*, or the Internet of Things, based on the connectivity and interactivity between people, information, processes and objects, through technologies that provide access to the network by anyone, anywhere, anytime, using any devices – including multifunctional devices with smart sensors (such as appliances, cars, clothing, etc.), as well as applications that dynamically adapt to people's needs (Davis 2012).

According to this paradigm, objects embedded with sensors become a source of information, with the ability to communicate (McKinsey & Company 2010). *Ubiquitous Computing* was defined for the first time by Weyser in 1993. According to Weiser et al. (1999) *Ubiquitous Computing* could be defined as the "physical world that is richly and invisibly interwoven with sensors, actuators, displays, and computational elements, embedded seamlessly in the everyday objects of our lives, and connected through a continuous network".

The term Internet of Things was first used by Ashton (2009) in 1999 (Atzori et al. 2012), in a presentation that linked usage of RFID (Radio Frequency Identification) to traceability in the supply chain of the consumer goods industry.

Since then, IoT has been interpret by some authors as a technology so innovative that its impact on society will be equivalent to that of an Industrial Revolution (O'Brien 2016).

3   IoT in Education

This section presents an overview of IoT in education, presenting some examples already experienced as well as a discussion about potentials and challenges.

3.1   An Overview

In the NMC Horizon Report of 2012 (Johnson et al. 2012) the application of IoT is mentioned, for the first time, as a future long-term trend (four to five years) in school adoption. This report refers to IoT as the next step in the evolution of intelligent objects where the boundary between the physical object and the digital information is blurr by its interconnectivity. Their relevance to teaching and learning is refer as the ability to attach small electronic devices to any object in a discrete way and use them to track, monitor, maintain and record data on that same object. This 2012 report reinforces the idea that the school (institution) can use these devices to monitor, track and inventory their facilities and objects, as well as granting automatic access authorizations for students, teachers and other staff members to certain locations.

Subsequently, the NMC Horizon Report 2015 (Johnson et al. 2015) returns to the topic and classifies IoT as an important development of educational technology for Higher Education, in a long-term perspective. Its potential use in teaching and learning is finally gaining prominence, especially through the concept of "hypersituation", as IoT's benefit in education. "Hypersituation" is the ability to amplify knowledge based on the user's location, contextualizing it from its geolocation, i.e., students carrying mobile devices that can collect immense interdisciplinary information emitted from the surrounding area. For example, a student exploring a city's historic center can search its surrounding environment through an architectural, political, or biological feature, depending on the information sent by the emitting devices of that urban setting and his or her interests at that time. Along the same lines, Cisco Systems envisions IoT as a context-sensitive environment where objects can communicate with the student and vice-versa to generate interactive learning experiences (Selinger et al. 2013). In this scenario, students will be able to monitor their own surrounding environment and collect real-time data for future study, with data emitted by these connected environments (Johnson et al. 2015).

The NMC Horizon Report 2015 (Johnson et al. 2015) also refers that, as the understanding of this emerging technology is increasing, universities are trying to take advantage of this opportunity to give their students more insight into the power of IoT. As an example, it is mentioned a project, developed by a consortium of four universities in partnership with an electric car manufacturer and a network research institution, to promote sustainable practices and initiatives to support energy efficiency. In this project, data sensors are used in vehicles to investigate various issues related to the effectiveness of public transport, psychological effects on drivers and gamification.

According to Benson (2016), IoT has the potential to produce significant gains in higher education institutions, especially in the automation of buildings, energy management, maintenance systems, access systems to buildings and spaces, environmental control, large systems of research environments, academic learning systems and safety systems for students, teachers, staff and the public.

3.2 Potential

The emergence of IoT ecosystems in schools will provide help for teachers, students and entrepreneurs to share various types of data in an open way. Teachers and students will have the opportunity to measure and share data with IoT technologies in a way that promotes diversity in the learning process and enables students to investigate and address real-world challenges using data made available by their environment. In order to awaken the creative role of students using IoT, it is necessary to build a social and technical ecosystem that integrates hardware, data, associated content and services. This ecosystem should provide easy access to information, assist interpretation of this data and empower students to act on their own interpretations (Joyce et al. 2014).

The use of IoT with this purpose allows the student to provide information that is in context with their age, interests and geographical location. It makes possible to address real and concrete challenges from the surrounding environment, making it easier for the students to construct their own knowledge (Constructivist approach) and the adoption of *flipped classroom* and multidisciplinary methodology. Considering this reality, IoT can promote that the contents treated in the class have a greater cultural and social affinity with the student.

An allusion is made here to the *flipped classroom* methodology, because with IoT it will be easier for students to assume a more active role in searching, sharing and processing of data, where the teacher will have mainly a guiding role.

Regarding multidisciplinarity, it is related to the fact that there is a greater possibility to develop competences of other curricular areas, on an interdisciplinary approach, as exemplified in (Ministério da Educação 2004), that reports the case of students dealing with specific data and, simultaneously, developing one of the competences required by the National Program for Basic Education.

IoT has the potential to contribute for the classroom to become an "open" space, where the physical limitations will not be relevant to the interpretation of the outside world, which can be monitored, analyzed and studied in real time, creating "hypersituation" provisions.

Regarding the school manual, this will not only be an object, but a dynamic service, which is constantly update taking into account different variables such as the age, interests and location of its users. Furthermore, it will take on a strong role as a guider in access to the information available on the network.

3.3   Challenges

Per several authors (Selinger et al. 2013; Dhungel 2015; Sundamaker et al. 2016; Harris 2016), some main challenges have arisen for the introduction of IoT in education. These challenges are related to security and privacy, especially of students, because of their learning history, personal details and even location.

Other challenge is about data storage, because IoT allows access to large amounts of data with the possibility of being constantly updated and there are need to storage it. This and the implementation (or adjust) of resources (hardware, software, and scholar equipment's) have costs.

Besides this, it is necessary develop training for teachers, since they are main agents of change, mobilizers of new educational practices and responsible for setting main guiding principles for the effective use of these new technologies. And in parallel it is necessary to develop didactic resources so that these teachers can employ them in their classes.

3.4   Some Examples

In this subsection, we present some projects and initiatives that involve the use of IoT in an educational context, at the level of basic education (Table 1).

4   Smartbook

As the school textbook is one of the resources still most used by teachers, namely in Portugal, it is important to reflect the impact that IoT technology may have on it. Today, school textbooks have become more inclusive and interactive, taking over not just as a resource but also as a service. The IoT, due to its potential, will allow the acceleration of this process, leading to a greater democratization in the access to

Table 1. Some projects and initiatives that involve the use of IoT in educational context

Institution	Project	Description
PTC academic program	Academic program, K12 program and IoT institute	A PTC (software company) initiative that aims to train and equip teachers with IoT knowledge – providing software, communication platforms and complete tutorials
The Internet of (School) Things	iotschool.org	The purpose of this UK project is to teach students and teachers how to measure and share data from various sources (such as air and soil temperature, humidity, among others, to encourage students to experiment with plant breeding)
ThingLearn	www.thinglearn.com	North American project aimed at sharing resources to help and encourage teachers to carry out activities involving IoT
Bosch – Bosch IoT Lab	Room Climate Monitoring Systema	Project to monitor the environment of a classroom. In other words, with IoT, the air quality of the classroom is evaluated, if it is not within the desired parameters, a picture with the photograph of the scientist Albert Einstein turns green. Whenever this happens the teacher must open the window so that fresh air enters the room and the painting returns to its normal color
Portuguese association of teachers of informatics (ANPRI)	Workshops	ANPRI has been carrying out several workshops across the country in the IoT area, especially on the Arduino platform. They have also presented conferences on the subject, with the aim of sensitizing teachers to IoT
IoT Digi Class	IoT Digiclass project	European project involving European companies and schools, including one from Portugal (in Baião), which aims to improve the integration of IoT ecosystem systems in schools. Includes several modules: the cafeteria, the classroom, the school concierge and the library

information, which will be in constant updating and adaptation to the reality of the student. The incorporation of IoT in the textbooks generates, for example, the possibility of exploring, in an articulated way, other paradigms of interaction, such as Augmented Reality or mobile computing, enriching the educational experience of the book.

In terms of functionalities, it will be possible for the student to solve exercises using real data that will be constantly update, and the manual will be taken as a means of communication with the whole school community. It can be consider that the local community feeds the manual itself, providing the actual data of the surrounding environment: geography, climate, industry, economy, population, etc. It is the "hypersituation" in action in the concrete and geographical context of the student.

Through the manual, the student can also access groups with similar interests, as the selection of interest in themes will be crosscheck with the entire community database. This may contribute to the personalization of education, enabling students to have access to more information in their personal areas of interest.

As a service, the textbook can notify the student to take breaks (for example), give suggestions on possible research fields or interest groups, or create concept maps. It may serve as an identification and tracking device, providing information about its owner to both the school institution and the parents.

Regarding contents, these may be constantly update according to the interest, age and geographic location of the student. Through the manuals, students can access nature data in real time allowing an observation and analysis of the environment without the need for travelling. The manual incorporates and makes this information available to the student, reflecting the actual conditions of the surrounding environment, allowing a greater contextualized awareness and reduction of the level of content abstraction.

Teachers will have the possibility to check what contents their students have more difficulties with and can then help them by adding notes, exercises or referring to electronic addresses with relevant information. They may also communicate directly with the student or carry out evaluation moments monitored by themselves.

5 Data from Sensors and Open Data

One relevant question that arises is how can we get the data to "empower" smartbooks and keep them constantly updated as referred to in the previous section? In relation to this, we envision two ways of reaching this objective: through sensors that will collect data from the environments; and through open data.

According to the Portuguese Agency for Administrative Modernization (Agência para a Modernização Científica 2016) *open data* is data that can be freely used, reused and redistributed by anyone, i.e., data that can be used without any kind of restriction.

Huijboom and Van de Broek (2011) argue that the publication of government data can enable citizens to exercise their democratic rights. However, *open data* can also be consider as an excellent resource for teaching. This is because this data in educational contexts may enable students to develop argumentative discourse (Weinberger and Fischer 2006). To establish objectives in research-based activities, in which students learn data organization models and academic content (Mazón et al. 2014) and to carry out activities based on collaboration, analysis of information and data, communication of results and relate them to specific scientific or social problems (Fisher et al. 2007) in an interdisciplinary way.

It is well establish that the use of real and local situations in learning activities, when approaching real-world contents, increases the motivation and the commitment of the students, thus improving learning. *Open data* can be an asset as it can be easily use in the development of personalized learning activities in order to respond to the learning objectives and in which the teacher has the feedback of the progress of the students that may even lead, if necessary, to the adaptation of planning (Chui and Farrell).

The use of *open data* in educational activities allows, in addition to specific content learning, the teachers to incorporate a civic education component into their classes, which complements the curriculum.

In this sense, for example, the use of sensors (such as temperature, humidity, pH, etc.) to monitor a greenhouse in real time provide data that could be then also used when exploring contents related to Physics, Chemistry, Biology, Mathematics and Geography.

Another way of obtaining data in real time, for the exploitation of the contents of these disciplines, would be the use of open data, which, for the sake of transparency and community service, an increasing number of institutions are making available. Data about the same parameters as the later example could be obtain through a data feed from the World Weather Online site, from NASA, from ESA or from other national institutions, such as in Portugal the National Hydrographic Institute.

Table 2 identifies a set of currently available open data feeds that may be used in learning contexts.

The combination and comparison between data obtained by sensors and open data would also allow students to assess the quality of the data and improve the scientific analysis of the facts.

In this way, the use of these data sets deployed through feeds would be an asset to a meaningful learning experience, in which the contents that are explored could be contextualized resulting in more motivating learning situations for the students.

6   Final Remarks

In recent years, the trend for IoT has been growing, both in terms of connected devices, and in terms of transactions and adherence by several institutions – particularly educational institutions, as discussed previously. With the almost exponential growth of IoT in the electronic market and its potential explored in the course of this article, it is practically impossible not to imagine its entry into the educational system, where its surplus value is evident.

Most of the projects, made public, aimed at integrating IoT into education were developed under a perspective of optimization and management of infrastructures (mapping of equipment, energy efficiency, comfort in the classroom, etc.) and safety (management of entries, etc.). However, the usage of IoT in learning is still in its early age.

It is assumed that the potential of IoT in Education is enormous, especially with the use and implementation of the concept of "hypersituation", that is, the amplification of information generated in real time by multiple sources namely embedded in the

Table 2. Examples of open data feeds that may be used in learning contexts

Institution	Url	Brief description of the data available
NASA	https://www.nasa.gov/content/nasa-rss-feeds	Contains headlines, summaries and links to full content on NASA Web sites
World weather online	https://www.worldweatheronline.com/porto-weather/porto/pt.aspx	Allows access current, past and future weather data for use in apps and on websites. The weather API provides local weather, ski and mountain weather and marine, sailing and surfing data. Weather can be searched for using a variety of information, including postcode, latitude and longitude. The weather API is easy to use and delivers advanced, in-depth weather information
ESA european space agency	http://www.esa.int/por/ESA_in_your_country/Portugal/RSS_Feed	Provides information from your top news to updates on various specific programs and projects
Instituto Português do Mar e da Atmosfera	https://www.ipma.pt/pt/produtoseservicos/index.jsp?page=rss.xml	It provides information on seismic activity and weather forecast for more than 300 locations in Portugal
World air quality index	http://aqicn.org/map/world/pt/#@g/-0.3048/146.9531/0z	Provides air quality data in the world
British geological survey - natural environment research council	http://www.earthquakes.bgs.ac.uk/earthquakes/feeds.html	Provides geological data
USGS earthquake hazards program	https://earthquake.usgs.gov/earthquakes/feed/	Provides automated notifications when earthquakes happen
Road traffic information in Portugal	http://services.sapo.pt/Metadata/Service/Traffic?culture=PT	Provides traffic information in Portuguese roads
European data portal	https://www.europeandataportal.eu/	Harvests the metadata of public data made available across Europe
STCP porto public transportation company	http://www.stcp.pt/pt/viajar/horarios-tempo-real/	Provides real time schedules of the public transportation system of the Porto city

(*continued*)

Table 2. (*continued*)

Institution	Url	Brief description of the data available
Flight aware	http://pt.flightaware.com/ commercial/flightxml/ documentation2.rvt	Allows to get information about scheduled flights, flights departing, flights that are on the way to the airport and flights that arrive at the airport
Quora	https://www.quora.com/ Where-can-I-find-public-or- free-real-time-or-streaming- data-sources	More info about open data feeds

geographic location of the student, providing a framework for contextualized learning supported on the surrounding environment.

In order for the IoT to be accepted and introduced into the education system, special attention and investment must be devoted to developing the perception and preparation by politicians, educators and society in general, raising awareness to its the advantages and challenges, as well as the need to ensure connectivity and continuous access to network and technology (Selinger et al. 2013).

In conclusion, we present the perspective of Marcel Bullinga (Pew Research Center 2012), who states that in the future education will be less focus on knowledge of facts as these will be easily accessible over the internet, and that children will learn less, but will be able to learn more.

References

Agência para a Modernização Científica (2016) Guia Dados Abertos – AMA|Dados Gov.com

Ashton K (2009) That "Internet of Things" thing. RFiD J 22(7):97–114 http://www.itrco.jp/libraries/RFIDjournal-That.Internet.of.Things.Thing.pdf

Atzori L, Iera A, Morabito G (2010) The Internet of Things: a survey. Comput Netw 54 (15):2787–2805. doi:10.1016/j.comnet.2010.05.010

Atzori L, Iera A, Morabito G, Nitti M (2012) The social Internet of Things (SIoT) – when social networks meet the Internet of Things: concept, architecture and network characterization. Comput Netw 56(16):3594–3608. doi:10.1016/j.comnet.2012.07.010

Benson C (2016) The Internet of Things, IoT systems, and higher education. EDUCAUSE Rev 51(4):6   http://er.educause.edu/articles/2016/6/the-internet-of-things-iot-systems-and-higher-education

Botta A, Donato W, Persico V, Pescapé A (2016) Integration of cloud computing and Internet of Things: a survey. Future Gener Comput Syst 56:684–700. doi:10.1016/j.future.2015.09.021

Chui M, e Farrell, D (nd) A closer look at open data: opportunities for impact. Government designed for new times

Davis M (2012) Semantic wave 2008 report: industry roadmap to Web 3.0 and multibillion dollar market opportunities. Executive summary

Dhungel R (2015) The evolving challenges of IoT: exploring higher education. In: IBM Big Data and Analytics Hub

Fischer G, Rohde M, Wulf V (2007) Community-based learning: the core competency of residential, research-based universities. Int J Comput Support Collab Learn 2:9–40

Gubbi J, Buyya R, Marusic S, Palaniswami M (2013) Internet of Things (IoT): a vision, architectural elements, and future directions. Future Gener Comput Syst 29(7):1645–1660. doi:10.1016/j.future.2013.01.010

Harris J (2016) IoT can revolutionize education, but challenges must be addressed. In: Remo Software

Huijboom N, Van de Broek T (2011) Open data: an international comparison of strategies. European J ePractice 12:4–16

Johnson L, Adams Becker S, Estrada V, Freeman A (2015) NMC horizon report: 2015 higher education edition, Austin, Texas

Johnson L, Adams S, Cummins M (2012) The NMC horizon report: 2012 higher education edition, Austin, Texas

Joyce C, Pham H, Fraser DS, Payne S, Crellin D, McDougall S (2014) Building an internet of school things ecosystem-a national collaborative experience. In: ACM International Conference Proceeding Series, pp 289–292. doi:10.1145/2593968.2610474

Mazón JN, Lloret E, Gómez E, Aguilar A, Mingot I, Pérez E, Quereda L (2014) Reusing open data for learning database design. In: Computers in Education (SIIE), International Symposium

McKinsey & Company (2010) Innovation and commercialization, 2010: McKinsey Global Survey results

Ministério da Educação – Departamento de Educação Básica (2004) Organização Curricular e Programas do Ensino Básico -1.º Ciclo. 4ª Edição. MEDEB, Lisboa

Mukhopadhyay SC, Suryadevara NK (2014) Internet of Things: challenges and opportunities. In: Mukhopadhyay SC (ed) Internet of Things: challenges and opportunities. Springer International Publishing, pp 1–17. doi:10.1007/978-3-319-04223-7_1

O'Brien HM (2016) The Internet of Things. J Internet Law 19(12):1–20

Perera C, Liu CH, and Jayawardena S (2015) The emerging Internet of Things marketplace from an industrial perspective: a survey. IEEE Trans Emerg Top Comput 3(4):585–598. doi:10.1109/TETC.2015.2390034

Perera C, Zaslavsky A, Christen P, Georgakopoulos D (2014) Context aware computing for the Internet of Things: a survey. IEEE Commun Surv Tutor 16(1):414–454. doi:10.1109/SURV.2013.042313.00197

Pew Research Center (2012) The Internet of Things will thrive by 2025. http://www.pewinternet.org/2014/05/14/internet-of-things/. Accessed 12 Mar 2017

Ray S, Jin Y, Raychowdhury A (2016) The changing computing paradigm with Internet of Things: a tutorial introduction. IEEE Des Test 33(2):76–96. doi:10.1109/MDAT.2016.2526612

Selinger M, Sepulved A, Buchan J (2013) Education and the internet of everything: how ubiquitous connectedness can help transform pedagogy. White paper. Cisco Systems, San Diego

Sundamaker H, Verdouw C, Wolfert S, Freire L (2016) Internet of food and farm 2020. In: Vermessan O, Fries P (eds) Digitising the industry – Internet of Things connecting physical, digital and virtual worlds, pp 129–152

Sundmaeker H, Saint-Exupéry A (2010) Vision and challenges for realising the Internet of Things

Vermesen O, Friers P (2016) Digitasing the industry – Internet of Things connecting the physical, digital and virtual worlds. IERC - IoT European Research Cluster

Weinberger A, Fischer F (2006) A framework to analyze argumentative knowledge construction in computer-supported collaborative learning. Comput Educ 46:71–95

Weiser M, Gold R, Brown JS (1999) The origins of ubiquitous computing research at PARC in the late 1980s. IBM Syst J 38:693–696

Xia F, Yang LT, Wang L, Vinel A (2012) Internet of Things. Int J Commun Syst 25(9): 1101–1102

Participatory Evaluation as Starting Point to Design for Smarter Learning Ecosystems: The UTOV Case History

Carlo Giovannella[1,2(✉)]

[1] University of Rome Tor Vergata – Dip. SPFS, Roma, Italy
`carlo.giovannella@uniroma2.it`
[2] ASLERD, Rome, Italy

Abstract. In the recent past it has been shown that participatory evaluation constitutes an alternative approach to traditional benchmarking of learning ecosystems capable to provide an insight on the smartness of universities and schools. Such alternative approach has also been shown capable to make emerge problems and desiderata from the opinions of all players involved in the educational processes.

In this paper we report on the outcomes of a participatory evaluation process carried on along three years at the University of Rome Tor Vergata to show how one can use them to go beyond the evaluation stage and guide a bottom-up design process to foster the achievement of smarter learning ecosystems.

Keywords: Smart learning ecosystems · Participatory evaluation · Design for smart learning ecosystems

1 Introduction

In the recent past we have shown that usual approaches to University benchmarking, although useful to catch the interdependence between campuses and pertaining territories on macro socio-economical basis (Giovannella 2014), are not capable to intercept the criticalities of the local contexts and, as well, expectations of the players involved in the learning processes. In particular they fail in identifying the multidimensional wellbeing state of the students that, after all, are the main target of the learning processes.

To make emerge descriptions of the learning ecosystems closer to students' feelings and expectations - and when possible also to those of the other players involved in the learning process (teachers, technicians, parents, territorial stakeholders) - ASLERD (ASLERD 2015) has developed a bottom-up participatory evaluation procedure (Giovannella 2015) that has been tested and validated on several European Universities (Giovannella et al. 2015; Giovannella et al. 2016) on a three years period by now. ASLERD approach allows to perform a multidimensional evaluation of the learning ecosystems and to work out a set of numerical indicators and indices (see for example Table 1). These latter serve as basis to perform a principal component analysis (PCA) (Jolliffe 2002; Hotelling 1933) and compare the levels of smartness achieved by the campuses. Accordingly to this procedure, the overall smartness of the University of Tor Vergata appears to be quite limited when compared with most of the other campuses

Ó. Mealha et al. (eds.), *Citizen, Territory and Technologies: Smart Learning Contexts and Practices*, Smart Innovation, Systems and Technologies 80, DOI 10.1007/978-3-319-61322-2_7

(see Fig. 1) involved in the validation procedure, and did not changed substantially along the period of observation: three academic years (see Table 1).

Table 1. Mean values and standard deviations of the indices worked out from the answers given to the quantitative questions of the ASLERD questionnaire. The data refers to the academic years: 2014–2015, 2015–2016, 2016–2017.

Dimension	2014–2015	2015–2016	2016–2017
Infrastructure	5,86 ± 0,23	5,60 ± 0,23	5,96 ± 0,23
Food service	5,94 ± 0,22	6,23 ± 0,20	6,28 ± 0,19
Environment	6,35 ± 0,25	6,6 ± 0,20	6,87 ± 0,21
Access to admin service and info	5,91 ± 0,20	5,83 ± 0,20	5,95 ± 0,17
Mobility	6,40 ± 0,24	6,61 ± 0,20	7,14 ± 0,18
Safety	6,24 ± 0,26	6,88 ± 0,25	6,91 ± 0,21
Socialization support	5,28 ± 0,22	5,72 ± 0,20	5,99 ± 0,22
Challenge	5,38 ± 0,20	6,09 ± 0,19	6,23 ± 0,18
Satisfaction	6,85 ± 0,22	6,92 ± 0,18	7,21 ± 0,20
Self-fulfillment	6,98 ± 0,21	7,22 ± 0,17	7,41 ± 0,17

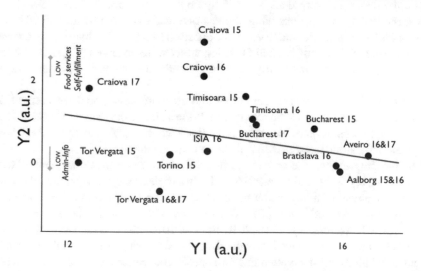

Fig. 1. Positioning of the universities on the plane identified by the two principal components, Y1 and Y2. Y1 and Y2 have been derived from a PCA applied to a reduced set of indices: Infrastructures, Food services, Access to admin service and info, Support to socialization, Challenges, Self-fulfillment selected as explained in (Giovannella et al. 2016). The red line indicates the direction of increasing "smartness" (from left to right).

Beside multiple choice questions useful to extract numerical indicators, ASLERD questionnaires contain also open questions aimed at making emerge criticalities and expectations on each one of the dimensions that contribute to define the level of smartness of a learning ecosystem (Giovannella et al. 2015). To have a better insight on students' feeling we decided to analyze the answers they gave to such open questions.

The outcomes have been used as input to start a bottom-up design process aimed first at engaging and empowering students to foster an support the campus evolution toward the achievement of a higher level of smartness.

In the following, thus, we present first the outcomes of the text analysis and, then, we show how they have been used as input of a design process intended to mitigate the problems emerged from the survey. Main goals of the design process have been the creation of new services to be used on mobile phone and the improvement of some of the existing ones. To conclude we describe the outcomes of the on-going participatory validation procedure of the proposed solutions.

2 Data Analysis

During last three academic years (a.y.) the ASLERD questionnaire for Campuses has been filled by a sample of students of the Tor Vergata University, mainly bachelor ones, attending curricula either in the scientific and in the humanistic domains. 81 students took part in the survey during the a.y., 2014–2015, 78 students in a.y. 2015–2016 and 80 in the present a.y. Overall, about 30% of the respondents were male and 70% female. In all three runs we observed a similar "fatiguing" effects: around 70% of the students answer the first open questions while the last one was answered by around 20% of the participants (see Fig. 2). All multiple choice questions, on the other hand, were answered by more than 90% and only a small fatiguing effect were observed. These are observations that guarantee the trustability of the collected data (see Ref. (Giovannella et al. 2016)).

The numerical outcomes (mean values and standard deviations) worked out from the multiple choice questions are reported in Table 1. It is worthwhile to stress that they stabilize – with oscillations of the values contained within 1–2% - any time 40–50 people of a given population answer ASLERD campus questionnaire.

The meaningful increase of some of the indicators (more than one standard deviation), between a.y. 2014–2015 and 2016–2017, highlighted in grey, is mainly determined by the involvement of a larger number of students attending humanistic courses with respect to those involved in a.y. 2014–2015 survey. If we compare the outcomes of the surveys carried on during a.y. 2015–2016 and a.y. 2016–2017 no meaningful variations can be detected and this is well explained by the similarity of the composition of the student populations. Students attending humanistic courses, in fact, seem to have a better feeling on some of the aspects that contribute to the determination of the overall smartness of the campus.

As stated above, the questionnaire includes also open questions that were formulated to make emerge explicitly the criticalities hidden behind numerical evaluations (indicators and indices) and to address all topics that contribute to the overall level of smartness of the campus and, thus, to the student wellbeing. Here below we report the outcomes of the text analysis that has been performed on the answers given by the students to the open questions.

It is worthwhile to stress that comparing the three data sets, the overall emerging landscape did not change substantially.

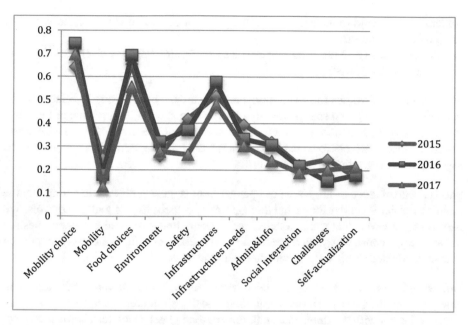

Fig. 2. Percentage of people who answered open questions as function of the order of presentation of the questions for academic year 2014–2015 (blue), 2015–2016 (red), 2016–2017 (green).

Mobility. On average more than 40% of the students use private transportation because the public transportation system is not considerate adequate: the suburbs are not fully covered and often multiple connections must be used by people living in there; often the transportations are crowded, the traveling time too long and service frequency too low. On the other hand public transportations are used to spare money and by people that have no alternatives. Cars are used because parking is not a problem at the University, because no time has to be lost in waiting for a bus and, overall, to feel free to move at any time. Bikes are, basically not used because biking in the city is considered very dangerous due to the heavy traffic and to the travelling speed of the cars.

Internal Mobility. Since the campus is one of the largest in Europe students are required to cover quite relevant distances to move between faculties and to go to the student canteen (this is particularly true for the students of the science faculty). No internal shuttle service is available and displacement with public transportation may be quite time consuming. In addition students claim that the indoor signage is quite outdated and this does not allow always to find classes, offices and professors' rooms. Similar problems about signage have been detected also in Aveiro (Galego et al. 2016).

Food services. A large number of students think that having a lunch at the student canteen is too time consuming either because of the time of the displacement and because of the queue that could be quite long, depending on the time slot chosen. Often the break they have at lunch time is not long enough. On the other hand, food sold by faculty bars is not considered to have an interesting quality/price ratio. Because of this most of the students try to avoid to have a lunch in the Campus and when strictly necessary they bring a lunch-box from home.

Similar problems have been detected also in Aveiro (Galego et al. 2016) and in Trento (De Angeli, 2014).

Environment. Two are the main students' concerns. The first one is related to the maintenance of the green spaces (mainly grass cutting and plants care) and of the urban furnishings (park benches, vases, etc.) The second one concerns the separate waste collection that seems not to be practiced in a satisfactory manner: often the bins for the separate waste collection are missing or full.

Safety. Surveillance of large area of open parking is quite problematic and in fact the main complaint is for car thefts and damages and, as well, for the lack of controls. We may note, however, that the percentage of students that answered the open question about safety is consistently decreased in a.y. 2016–2017, see Fig. 2, possibly to indicate that such criticality may have diminished with the time.

Infrastructures. Maintenance of the infrastructures is one of main criticality that emerges from the survey. The status of chairs, desks, air conditioners, etc. tends to get worst and worst with the time, while equipments tend to get obsolete. The wi-fi connectivity is slow and intermittent. In some cases the classrooms are to small and only few places for students to study or socialize are available. Often students have to go around to find for available spaces (ex. unoccupied classrooms). The lack of a sufficient number of spaces dedicated to student activities has been detected also in Aveiro (Galego et al. 2016).

Administrative and info services. Information relevant for students provided by the university website are considered incomplete or difficult to be found. Digitalization of the administrative services are not complete and often students are obliged to go to the student administrative offices with the risk to stay for quite a bit in the queue. Getting in touch with professors seems also quite complicated.

Social interaction. Concerns on the support offered by the University to the social interaction are focused on two main topics: (a) a quite limited number of cultural and sport activities; actually the university has no strong and organized student unions; the few existing associations are not interacting among them, promote only rare events and haven't reasonable aggregation spaces; (b) the limited and sporadic activities dedicated to job-linking, networking, a better knowledge of the territory and, more in general, the presentation of opportunities for the students. Criticalities listed in (b) have been detected also in Aveiro (Galego et al. 2016).

Challenges. Strictly correlated with the issues listed at the letter (b) of the previous topic (*social interaction*) are the complaints about the reduced number of challenges that involve students: projects, collaborations, lab activities, etc. It is very important to stress a trend that is very general for Italian learning ecosystems: students feel themselves under-challenged with respect to the competences that they think to have developed.

In fact they feel more *satisfied* and *self-fulfilled* than challenged.

Students, apart few exceptions do not complaint about the quality of the teaching but, rather on the possibility to transfer learning on practical activity and on the possibility to transform knowledge into skills and competences.

Considering all above, the emerging picture is that of a campus characterized by a limited capability to attract the students. Not by chance the main reason to choose the campus is its proximity to home. Immediately after comes the uniqueness of the curricula (chosen mainly by students involved in scientific curricula). Only about 25% have chosen the campus on the basis of the perceived quality of teaching and learning. At the end students seems to be quite happy with teaching and their personal development, but due to the criticalities described above *do not develop a sense of belonging*.

3   Design for Smarter Learning Ecosystem

To go beyond the evaluation phase and to empower students, the outcomes of the ASLERD questionnaire has been used to involve some students attending the degree in Media Science and Technology in a design process aimed at identifying solutions capable to mitigate the criticalities emerged and, thus, to help the campus to achieve a higher degree of smartness and student to increase their wellbeing. After a brainstorming activity focused on how to achieve, at least partially, the expected goals the focus has been concentrated on the design of an application for smartphone capable to offer a set of services, summarized here below.

Mobility. To mitigate the problems with external and internal mobility the app will offer a trusted car sharing service. An additional proposal is to offer also a bike sharing service for internal displacements but of course this is a service that would imply also a commitment by the University or by private investors. Another service offered by the app will be an interactive map of the campus that is expected to integrate also information on spaces (e.g. free classrooms) that could be used to carry on students' activities and the availability of professors and assistants.

Food services. The app will offer the possibility to see the menu of the week, to make her/his own choice and to reserve the preferred time slot to take a lunch, provided that the student has demonstrated her/his trustability. The aim is to mitigate the queue problem and, thus, reduce the time needed to have a meal. Another goal is also to optimize food consumption and preparation time with the hope to favor the increase of food variety and support the preparation of special food for those who need it. An app with similar functionalities, iFame, has been developed by the students of the Trento University (De Angeli 2014).

Environment. The interactive map offered by the app will provide information on the position of the collection points that allow to operate the separate waste collection and on the filling status of bins and dumpsters.

Another functionality will allow the students to report about maintenance problems (green space and urban furnishings) and to organize maintenance activities.

Safety. As for the environment section the app will allow to report on safety problems with the aim to develop a map (space and time) of the critical issues and support the University in organizing a better surveillance service.

Infrastructures. Again the app will allow to report on problems. In addition it will allow to monitor the time that will elapse from reports to troubleshooting by university. It will also allow students to organize, in special case, voluntary troubleshooting.

Administrative and info services. As for *Food services*, the app will allow to book appointments with student secretary and teachers and, as well with any other relevant structure offering services within the University. This to avoid queues and time wasting for students, but also to improve the efficiency of the University services.

Social interaction. The app will offer the possibility to organize events and activities and to look for the needed human resources. Moreover the app will propose a rewarding mechanism (credits) that can be used to get discounts, priorities or that can be exchanged to get help from other members of the community.

Challenges. The app will offer the possibility to announce or launch challenges either internal to the University and external, i.e. proposed by companies, associations, municipalities, etc. Also room to report on the outcomes of the participation in the proposed the challenges will be provided.

Overall the application aims at facilitating the constructive interaction among the players involved in all activities of the learning process in order to mitigate criticalities and solve problems, if possible in a participatory manner. It is also intended to support the development of a truly student community and to take as much as possible advantage of all the available opportunities that may derive from collaboration and cooperation.

Very simple rewarding mechanisms have been included to challenge students, provide advantages and strength relationships. Much more could be done on the gamification of the processes but it will be left to future revisions of the proposed application.

Before to start the development of a medium profile prototype we have put in place a validation process that was intended to involve more students. Preliminary outcomes of this on-going process are described in the following paragraph.

4   Design Validation and Conclusions

Taking into account the goals of the application we have prepared a new questionnaire to involve a consistent number of students in the validation of the proposed solutions. In Table 2 we report a summary of the questions that have been asked and the outcomes of the validation process. 125 students from all University faculties took part in the validation procedure, about 30% of them were male and 70% female. 100 over 125 were bachelor students.

Table 2. Outcomes of the validation questionnaire (depending on the formulation of the question the outcomes are reported as percentage or as mean values with the associated standard deviations.

Questions	Results
Mobility. How useful would be a student car sharing service?	$7,32 \pm 0,18$
Mobility. Would you share your car, in case you have one?	yes = 52%
Mobility. How useful would be a bike sharing service?	$6,77 \pm 0,21$
Mobility. How useful would be an interactive map to support internal mobility?	$8,20 \pm 0,18$
Mobility. Would you contribute to keep the map updated?	yes = 87%
Food service. How useful would be a service to select the menù and to book the time slot to take the lunch at the student canteen?	$7,34 \pm 0,21$
Environment. How useful would be an interactive map to support separate waste collection?	$5,84 \pm 0,24$
Environment. Would you contribute to keep the map updated about the filling levels of bins and dumpsters?	yes = 74%
Environment. Would you use an application to report on maintenance problems?	yes = 90%
Environment. Would you contribute as volunteer to keep clean and take care of the public area of the University?	yes = 61%
Safety. Would you use an application to report about safety problems?	yes = 93%
Safety. Would you contribute as volunteer to the surveillance of the public area of the University?	yes = 46%
Administrative and info services. How useful would be a service to book meeting with student secretary and more in general to avoid queues at the University's offices?	$8,51 \pm 0,15$
Challenge. Would you contribute to keep updated the information about challenges and opportunities for students?	yes = 70,4%
Social interaction. How useful would be an application to share with other students problems and solutions and to organize activities and events?	$7,23 \pm 0,20$
Rewarding mechanism. How much a rewarding mechanism could entice you to participate in the social activities referred to by the previous questions?	$7,68 \pm 0,19$
Student community. How much do you feel to belong to a student community?	$5,37 \pm 0,19$
Student community. How much do you perceive the existence of a student community?	$5,31 \pm 0,19$
Student community. Apart from helping in mitigating problems, how much an application offering the functionalities described in the previous questions would contribute to support the development of a student community at the University?	$7,39 \pm 0,16$

The outcomes of the questionnaire made emerge a strong attitude for problem reporting and a reasonable availability to contribute to keep information about the campus updated, but also a scarce propensity to be involved in social activities and in sharing her/his own car. An active university citizenship may certainly be encouraged by the activation of rewarding mechanism (reduced university taxes or other discounts).

However rewards could be not fully sufficient because of the absence of a cohesive student community and of a very low students' sense of belonging.

Despite of the lack of internal cohesion and sense of belonging, students think that the proposed application may: (a) drive the development of a student community at least to share, and find solutions to, common problems; (b) help to organize activities/events.

Getting into some more details:

Mobility. The amount of people that would be available for car sharing may considerably increase if the application will guarantee sharing of the expenses (fuel), and, overall, reliability of the passengers on several aspects.

Possibly the application should also allow to organize "meeting hours and corners" aimed at supporting car-sharing.

Bike sharing is considered a nice idea but too difficult to be implemented due to the lack of cycling paths and the dangerousness of the standard roads.

Food service. Students feel that an application will not be able to solve the problems of the queue at the canteen since they think that such service should be accompanied by a reorganization of the timetable of the lessons and, possibly, by the realization of a distributed system of students' canteens.

Environment and Safety. Active citizenship to take care of the common space is not considered a task for students. Moreover most of the students think that they have no time to contribute. Service should be provided by the University.

Social activity. A relevant part of the students tend to spend at the university the strictly needed time to attend lessons. The remaining part attempt to study where possible, since places in the studying area and libraries are limited. Only a very limited part of the students is interested in social activities also because of the lack of established and active student communities/unions.

We may observe that the lack of an established and cohesive student community doesn't make perceive the University as a place smart enough to be "lived", but rather as a physical space where one wishes to spend only the time needed to benefit of the services of interest. At a first superficial glance you get the impression that the students' well-being is strictly related to the efficiency and efficacy of services and opportunities provided by the University. At a second in depth scrutiny of the outcomes of the questionnaire, however, the desire to be part of a more cohesive student community emerges. Students, in fact, hope that the proposed application could help not only in reporting problems to have them mitigated but also in supporting the development of a student community.

The case study reported in this paper shows that ASLERD systemic and holistic bottom-up strategy to achieve a participatory evaluation approach represents also a significant starting point to activate student engagement. This is an important step forward with respect to the use of monitoring and evaluation procedures aimed only at benchmarking and at producing rankings, moreover it is also a step forward respect to procedures aimed only to produce a top-down action plans (regardless of the fact that the evaluation strategy could have been a bottom-up and participatory one). In the past a step toward the design for a Smart Campus have been done in (Galego et al. 2016) but the outcomes of the analysis were not used to activate students and engage them in a design process. Even before a very nice project, "Smart Campus – Creating services WITH and FOR people" (De Angeli et al. 2014), were lunched within the University of Trento with the aim to integrate smart campus and smart city applications to support not only academic life but also socialization and practical smart mobility. The initiative has been very successful in engaging students and citizens through user centered design methodology and hackathon but it was not sustained by the use of any "smartness" framework.

Many other universities are interested and involved in the development of the so called "smart campuses" but usually top-down design strategies are adopted and the focus is mainly on testing smart infrastructure (smart grid, etc.) as pilot for future implementations needed for the so called "smart cities". Since in all these cases the adjective "smart" is strictly related to the improvement of technological infrastructures and not to the wellbeing of the students (or other campus' players) they are not deemed relevant for this specific context.

For the future it would be very interesting to see if the process of engagement and empowering described in this case study would be adopted by the University of Tor Vergata as a "modus operandi" and, as well, if it could be transferred to other Universities or learning ecosystems like schools.

Considering the outcomes of (De Angeli et al. 2014), together with those of the validation questionnaire (see Table 2) it would be also very interesting to investigate further the role and relevance that a cohesive community may have in the development of a fully University citizenship and, thus, in driving an overall increase of the smartness of learning ecosystems.

Acknowledgments.   the author wish to thank all students that took part in the participatory evaluation and in the validation processes and in particular those that during the last three years collaborated in the development of small steps of the overall process: Beatrice Borghi, Christian Conti, Chiara Piseddu, Emiliano Tassone

References

ASLERD (2015). http://www.aslerd.org, https://en.wikipedia.org/wiki/ASLERD. Accessed Mar 2017

De Angeli A, Bordin S, Menéndez Blanco M (2014) Infrastructuring participatory development in information technology. In: Proceedings of the 13th participatory design conference, vol 1, pp 11–20

Galego D, Giovannella C, Mealha O (2016) An investigation of actors' differences in the perception of learning ecosystems' smartness: the case of University of Aveiro. IxD&A J 31:19–31

Giovannella C (2014) Where's the smartness of learning in smart territories? IxD&A J 22:59–67

Giovannella C (2015) Territorial smartness and the relevance of the learning ecosystems. In: ICS2 2015. IEEE Publisher, pp 1–5

Giovannella C, Andone D, Dascalu M, Popescu E, Rehm M, Roccasalva G (2015) Smartness of learning ecosystems and its bottom-up emergence in six European campuses. IxD&A J 27:79–92

Giovannella C, Andone D, Dascalu M, Popescu E, Rehm M, Mealha O (2016) Evaluating the resilience of the bottom-up method used to detect and benchmark the smartness of university campuses. In: ICS2 2016. IEEE Publisher, pp 341–345

Hotelling H (1933) Analysis of a complex of statistical variables into principal components. J Educ Psychol 24:417–441, 498–520

Jolliffe IT (2002) Principal component analysis, 2nd edn. Springer series in statistics. Springer, New York

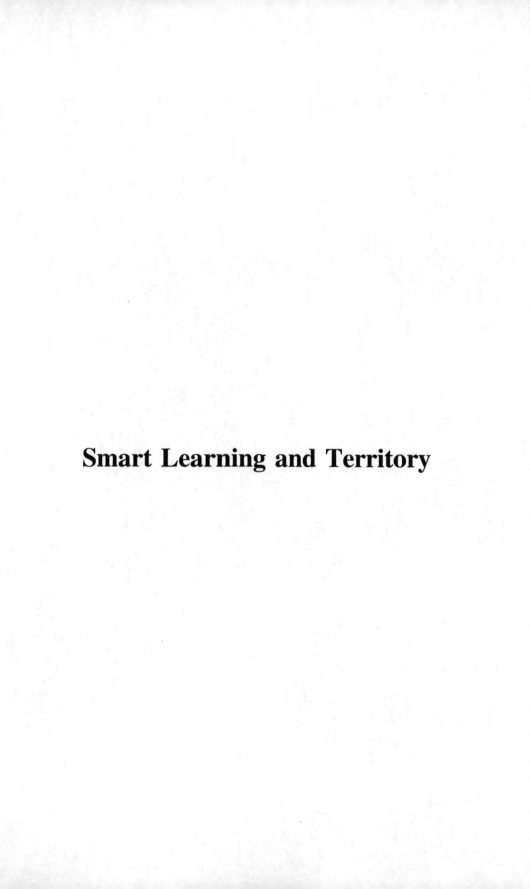

Smart Learning and Territory

DELTA: Promoting Young People Participation in Urban Planning

Mikael Reiersølmoen, Francesco Gianni[✉], and Monica Divitini

Department of Computer Science, Norwegian University of Science and Technology,
Trondheim, Norway
mikael.reiersolmoen@gmail.com, {francg,divitini}@ntnu.no

Abstract. Today urban areas are growing fast, in this process it remains a challenge to include the opinions of the public. This holds especially true for young people. From the 1960's, the rights of the children started to be recognized through the UN Convention on the Rights of the Child. Article 12 of the convention states that children should have the opportunity to express their views in matters that affect them, like urban planning. Our work started analysing the literature and providing an overview of participatory methods used to include children in land-use projects. We then reviewed existing mobile apps targeting participation in public life. Based on the findings from the literature a list of high level requirements was created to guide the design and implementation of the mobile app DELTA. The DELTA app support situated participation encouraging users to explore the urban environment, promoting awareness and critical thinking. Several user evaluations are performed during the development cycles, including expert evaluations, usability and field tests of the final prototype. Based on collected data and lessons learned, results are discussed in relation to participation and learning outcomes for children.

Keywords: Mobile technology · Participation · Urban planning

1 Introduction

Though public participation in urban planning and design of urban spaces is recognized as important, conventional methods of public participation like public hearings, questionnaires and committee groups have failed to engage the majority of the public (Roberts 2004; Irvin and Stansbury 2004). This resulted in the ongoing multidisciplinary search for more collaborative ways of participation, where opinions and knowledge of citizens and stakeholders are taken into account through authentic dialogs, building social capital and trust (Innes and Booher 2004).

This holds especially true for young people. Since the 1960's, the UN Convention on the Rights of the Child, Article 12 states that children should have the opportunity to express their views in matters that affect them, like urban planning. In Norway, the law regulating planning and building underlines the importance of involving young people and assigns to municipalities the responsibility to ensure their active participation (Norwegian Government 2008).

© Springer International Publishing AG 2018
Ó. Mealha et al. (eds.), *Citizen, Territory and Technologies: Smart Learning Contexts and Practices*,
Smart Innovation, Systems and Technologies 80, DOI 10.1007/978-3-319-61322-2_8

Involving young people in processes of urban planning is important not only because it is their right (Francis 1988), but also because of their ability to evaluate their environments and come up with ideas of their own (Gleeson and Sipe 2006; Bartlett 2002; Horelli 1997). In addition, the process of participation is also a process of learning, improving their "environmental awareness, knowledge and skills" (Wilks and Rudner 2013) and teaching them how to negotiate and respect other people's views (Corsi 2002).

Despite the increased recognition of young people as fully legitimated actors, it remains a challenge to sufficiently include them (Lansdown 2010). Attempts to do so can often be described as mostly including them for symbolic purposes and not supporting meaningful participation. To fully engage young people, there should be an effort to create new, innovative methods of participation that suit them, utilizes their knowledge and ideas, and make sure that their contributions are used.

In this paper we present the design and evaluation of DELTA[1], a mobile app to promote youth participation in urban planning. More specifically, we focus on (i) how situated engagement increases awareness of challenges and opportunities in the local environment and (ii) how game elements can promote the process of participation. Situated engagement means that participation happens at the specific physical location through immersive experiences (Korn 2013). This might lead to a more sustainable form of engagement (Gordon et al. 2011). The rationale for this design choice is connected to the complexity of urban planning, and to the fact that methods of participation should match the capabilities and interests of participants. In addition to facilitating participation, situated engagement might increase participants awareness of challenges and opportunities connected to specific planning projects. Urban planning is not only complex, it is often also perceived as something uninteresting by many young people. It is therefore critical to create a solution that engages them, possibly through game elements.

The paper is organized as following. Next section discusses different approaches to participation of young people in urban planning, followed by related work. Section 4 presents DELTA, with focus on main functionalities, while Sect. 5 describes the three main iterations of its design and evaluation. Section 6 summarizes the results and Sect. 7 concludes the paper and presents future work.

2   Participation of Young People

Despite increased awareness, young people are still today insufficiently included in urban planning (Haider 2007; Wilks and Rudner 2013). Conventional methods of public participation fail to engage them, often being characterized by a high threshold to participation. To overcome these challenges, new forms of participation must be developed. Hart (Hart 1992) in the attempt to stimulate a dialog around the topic, proposed a "ladder of participation" to help framing children's participation, ranging from a mainly symbolic participation to actual participation started by young people themselves. Frank (2006) reviewed several case studies about participation projects where children

---

[1] "delta" is the Norwegian word for "participation".

assessed their local environment, formulated plans to take action, and took steps to support implementation of their idea. The findings strongly emphasize the positive outcomes of such projects, and thus the importance of working towards inclusion of young people in planning processes. Lessons learned point in the direction of *"giving youth responsibility and voice, building youth capacity, encouraging youthful styles of working, involving adults throughout the process, and adapting the sociopolitical context"* (Frank 2006, 633).

As part of our work, we reviewed 14 studies addressing the challenge of young people participation (Reiersølmoen 2016). These studies were identified through a systematic mapping of the literature, following the methods described in (Keele 2007). Articles were collected from several online databases and screened in two iterations. Several studies reported inclusion of children that was rather limited and mostly for symbolic purposes, as in (Breitbart 1995; Cunningham et al. 2003; Mallan and Greenaway 2011; Passon et al. 2008). Most of the analysed projects are covering only the lower levels of the ladder of participation proposed by Hart. Other studies focus on the general lack of participation among young people (Laughlin and Johnson 2011; Torres 2012; Dennis 2006; Malone 2013). Several of the analysed studies recognize that the methods used to include adults in planning processes, like public hearings and consultation meetings, are not likely to provide good results when used with young people (Spier 2013; Breitbart 1995; Cunningham et al. 2003; Laughlin and Johnson 2011; Dennis 2006; Mallan and Greenaway 2011). This awareness led to the development of less conventional participation methods, including free drawing, walking tours, photography, specific drawing tasks, and artworks. Horelli (Horelli 1997) conceptualize participation methods into five different groups. Most projects utilized multiple methods, mainly aiming at evaluating an environment or letting participants express their ideas. From the review, it is clear that the most popular participation methods are highly situated, as for example Lawrence Halprin's concept of walkshop that combines walking tours with specific tasks along the way. Walkshops *"are based on the idea of experience, interaction and communication, not just talking. They become more profound because the approach knocks out the usual seminar or lecturing process that gets in the way of most creativity, because it informs people rather than allow them to discover through personal experience"* (Halprin et al. 1999, 43). In a similar vein, other methods let participants walk free in the physical space, while using drawings or photography as a way to focus and describe the environment. In some cases, participants walk together, sharing thoughts with each other and with facilitators, enabling for negotiation and sharing of ideas. Summarizing our review, there is a need for new forms of- and opportunities for participation to be developed, as most young people have yet to experience being actively involved in planning processes that affect their lives (Lansdown 2010). The reviewed projects show different efforts to promote real inclusion using methods that engage young people. Incorporating situated action in participation methods seems the most promising approach, and is backed by research stating that young people get an understanding of the environment by using their senses (Haider 2007) and when they engage with its features (Chawla and Heft 2002). These methods are, however, costly and time-consuming, generally including only a small number of participants. In the next section we explore how mobile applications have been used to support participation.

3   Related Works

Given the situated nature of many of the participation methods identified in the literature review, mobile apps seem to be a promising approach for technological support. The potential of mobile participation also emerged from interviews with city officials and political decision makers in urban planning, who preferred a web-based mobile solution over other approaches such as interactive public screens and design tables for multiple users (Oksman et al. 2014). Participation via smartphones is also expected to be more appealing for young people who normally do not interact with government services (Clark et al. 2013), as they are early to adopt new technologies and shape how they are used (Castells et al. 2009). Example of mobile apps for public participation include solutions that enable citizens to report damages in the city, or provide up-to-date information about it. Most of the efforts to facilitate participation using smartphones fall under the category of informing apps, there is a lack of apps that enable the public to participate in more profound ways (Bohøj et al. 2011). We reviewed the literature for apps that facilitate situated engagement in urban planning. The features and characteristics of the five apps that were analysed are now briefly presented.

Mobile Democracy (Bohøj et al. 2011) allows citizens to express agreement/disagreement, comment and upload photos related to particular topics discovered by browsing a map. Augmented reality is also used to position urban elements on top of the smartpthone camera view. In **Augmented Reality** (Allen et al. 2011) the focus of the app is to overlay predefined graphic models on existing buildings. The user could therefore only tilt and pan the phone to look at the model, but not move around. It is also possible to rate the different models. **FlashPoll** (Schröder 2014) is an app allowing users to answer location-specific polls, it's designed to overcome the shortcomings of face-to-face participation. Polls are location-based and citizens can participate only when they move in the poll area. **Tienoo** (Kangas et al. 2015) was developed to collect location-specific opinions about forests in Finland. It allows geolocated data collection to happen in real time.

Community Circles (Lehner et al. 2014) is based on ideas coming from location-based games, the goal is to enable long-term participation in urban planning. Users create location-specific issues, ideas, opinions or polls that other users can comment, upvote or downvote.

4   DELTA: An App for Participation

In this section we present DELTA, the app we designed to promote participation in urban planning. While in this section we focus on the main functionalities, in Sect. 5 we present the design process and how the app developed through three main iterations. DELTA is designed to support situated participation and it is inspired by the idea of walkshops (Halprin et al. 1999). DELTA has as target group young people with fully developed writing and reading skills, mainly teenagers.

4.1 High level requirements

After reviewing existing apps and participation methods, we identified a list of high-level requirements for our solution.

HLR1 Location-based: DELTA should support activities to take place at the location where participation is desired, in terms of ideas or feedback on possible solutions. Participation should be based on participants actual perception of the place.

HLR2 Engaging: One of the implicit goals of DELTA is to include the voices of young people in urban planning, and therefore the user must be given an incentive to use the application, it must be engaging. One place to look for inspiration is geocaching. Common motivations for geocaching are: social walking, exploring new places, collecting caches, social status online, competition and challenges (O'Hara 2008).

HLR3 Collaborative: DELTA should support collaboration among users. Collaboration helps create better ideas, represents the importance of negotiation in urban planning and creates opportunities to share opinions.

HLR4 Project support: DELTA should inform the user about ongoing or planned projects in the city.

These requirements focus on the perspective of the citizens, without considering urban planners. Indeed, the app is complemented by a back-end solution for urban planners to add and manage projects, but this is outside the scope of the paper and it will not be explained further.

4.2 App description

DELTA is designed to enable and motivate young people to participate in urban planning. From the app, users can get an overview of active urban planning projects in their area and they are allowed to contribute in several ways: (i) they can complete surveys, (ii) post suggestions, and (iii) discuss suggestions. These functionalities are paired with game elements such as personal points (score) and achievements. The decision to structure the app around projects is related to HLR4. The inclusion of surveys is inspired by FlashPoll and Tienoo apps, described in chapter 3, while how the surveys are designed is largely inspired by the concept of walkshops, which relates to HLR1 and HLR2. The ability to post and discuss suggestions satisfies HLR3, while game elements relate to HLR2, which is connected to make the participation engaging.

The main screen of the app is a map showing the current location of the user. Locations of planning projects that are under development are marked on the map, with colors indicating one of three different states: no active survey, completed active survey, and uncompleted active survey. Users can select a project to read more about it and start the connected survey.

Projects - Projects should be added by planners, every survey is connected to and resolves around one specific project. Suggestions posted by users must likewise be posted within one of the active projects (Fig. 1, left). Exceptions are scores and achievements, explained more in detail later in the chapter, which are not project-specific. The app shows the location of the planning projects on a map so users get an overview of what planning projects are currently under development in the area.

Fig. 1. List of active projects, task answering interface and personal profile.

Survey - A survey is designed like a treasure hunt, where each task of the survey (Fig. 1, center) is connected to locations that the user has to find. To complete a task the user must be in close proximity of its location. The user will not see the coordinates marked on the map, but will see approximately how far away the location is. After a task is completed the user will have to navigate to the next with the help of a textual description which references the map and elements of the surrounding environment. The idea behind the surveys is inspired by Lawrence Halprin's concept of city walks. Participants receive a set of instructions that guide them to certain locations, where they are asked to contribute with ideas and thoughts about the place (Halprin et al. 1999). The walk is designed to bring awareness of problems and opportunities in the city (Hirsch 2011). The intention is not simply getting from one place to another as quickly as possible, but to see the connection between places (Hirsch 2011). Participant are forced to look up from the smartphone, increasing awareness of the surrounding environment. Game elements are used to create a more challenging and engaging participation process, without reaching a difficulty that can interfere with the experience. Participants are encouraged to cooperate, since the treasure hunt is likely an activity that people will enjoy doing together. Planners design the route they want the participants to follow, which allow to craft an experience that pays particular attention to specific places along the way. There are four different types of tasks, which consists of one or more questions. The difference in the task types is how the questions can be answered: it can be through a linear scale, multiple choice, check-boxes or free text.

Suggestions - A suggestion consists of an image, a title and some describing text. Other users can agree or disagree, and they can comment on it. Creation and interaction with suggestions can happen later in time, regardless of the physical position of the user.

The opportunity for users to comment and express agreement or disagreement on other user's suggestions is the fundamental way in which the app supports collaboration among users, as it encourages dialog between citizens.

Profile page - Each user has a profile page (Fig. 1, right) which shows an overview of the activities in the app: suggestions, comments, agreements and disagreements, and number of surveys answered. The profile page also shows the user's score and list of achievements.

Game elements - Certain activities are rewarded with points and achievements. As an example users get points for completing a survey, post suggestions or receiving agreement on posted suggestions. Achievements are rewarded when reaching predefined milestones, for example taking two surveys or after the fifth suggestion posted. A public leaderboard reports the ranking of the users. The game elements are intended to increase engagement using rewards and competition. This encourages high quality contributions since the best way to get a high score is to post suggestions that receive many agreements.

5   Design Process and Evaluation

The prototype described in Sect. 4 was developed through three iterations. The first iteration defined the concept, a low-fidelity prototype was created and evaluated using expert interviews. In the second iteration a functional prototype was developed and its usability tested. On the third iteration the prototype was refined and another expert evaluation was conducted. More information is available in (Reiersølmoen 2016).

5.1   Iteration 1: Expert evaluation

A first interactive mockup for internal use was built using an online diagram tool[2]. The mockup was then used as a blueprint for the first prototype of the Android app that was tested during the expert evaluation. The app allowed to navigate into different views, but the content displayed was static. The evaluation of the first iteration consisted of semi-structured interviews with four persons, with the main goal to present the concept and collect feedback regarding its potential. The subject of the interviews were: F1, 15 years old, member of the youth city council and representative of the end users of the app; F2, 28 years old and M1, 44 years old, respectively researcher and professor from the faculty of Architecture and Fine Arts, with expertise in urban planning; M2, 45 years old, employee in the local municipality with responsibility for involvement of young people. The interviews started showing some pictures of the app while in use in several urban contexts, then more detailed questions were asked while demonstrating different functionalities. Audio recording were collected and transcribed for analysis.

Results. F1 was positive about the concept, it was perceived as a tool to allow more people to participate, not only those who are actively engaged in committees. The ability to comment people suggestions was seen as a useful way to understand different

---

[2] www.lucidchart.com.

viewpoints. F1 believed that young people can be willing to perform the survey/treasure hunt on their own initiative if they are interested in the urban area. M2 reflected on how the approach of physically going in the area of interest to provide feedback might facilitate participation. It was also highlighted how different it is from the approach currently in use by the municipality. In comparison to other projects M2 was familiar with, DELTA was seen as more focused and connected to specific planning projects, which facilitated the planners in getting direct feedback and provided more guidance to the users. Discussing the requirement of being situated in the urban area, M2 thought that situatedness would increase awareness, especially if the area is not usually frequented by the user. However, he was unsure if a treasure hunt would make sense without a preliminary briefing with the users. Overall M2 believed that the tool *"can be valuable when facilitating children and youth's influence"* and the score mechanism can be an engaging factor when users compare their performance at the end of the day. M1 and F2 provided feedback more connected to the planner interface, that was desired as easy to use as the rest of the app. The fact that participants had to navigate using reference points in their surroundings, and not only an interactive map, was seen as a positive factor. M1 expressed some concerns regarding the motivation for using the app, and whether or not the game elements were enough to engage. Additional motivation could possibly come in the form of physical rewards along the treasure hunt. Another proposal to increase the engagement was to let people draw on top of the pictures added as suggestions, which was believed to be particularly fun for younger users.

5.2   Iteration 2: Usability evaluation

On the second iteration, the static data used in the first prototype was replaced by real data. A short pilot evaluation took place right after the functional prototype was ready, then the usability test was performed outdoor with a group of five university students coming from three different study programs. Participants were given first a short presentation of the concept, then they were requested to perform a set of tasks covering all the functionalities. The most comprehensive task required to complete a treasure hunt outside, around the university campus. Participants were encouraged to follow a think-aloud protocol. Data logging was performed following multiple strategies: (i) smartphone screen was recorded, including on-screen touch events; (ii) suggestions and notes were taken by the facilitator which acted as an external observer; (iii) users were equipped with a head-mounted camera, which allowed to capture how they interacted with the smartphone and the environment. At the end of the tasks, users compiled a questionnaire about the perceived usability of the app using the System Usability Scale (Tullis and Stetson 2004). In the analysis of the data we focused on identifying errors in the user interaction, defined as unintended or wrong actions the user made while completing a task. The footage from the head-mounted camera was used to extract the following metrics: (i) time spent on the survey; (ii) time spent finding locations; (iii) time spent answering tasks; (iv) time spent interacting with the app; (v) time spent interacting with the environment.

Results. We recorded 8 minor usability errors. In all the cases except one, the user expected a certain interaction, but nothing happened. Participants still managed to complete the given task in all case. Three more severe errors were reported, where the app resulted in an unwanted state. Only one of these errors was considered of high severity, based on how hard was for the user to recover from the unwanted state. During the survey, on average users spent 27% of their time answering the tasks and the remaining 73% finding the next one. In Fig. 2 we report statistics on the average time spent by the users interacting with the app versus the surrounding environment. Results from the questionnaires showed high usability (Tullis and Stetson 2004).

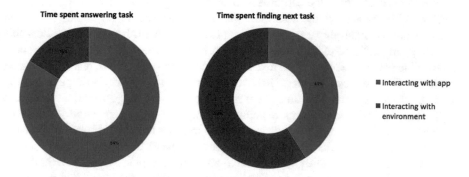

Fig. 2.   Time spent interacting with app and environment.

5.3   Iteration 3: Final expert evaluation

The most critical problems highlighted during the usability evaluation were fixed to make the prototype ready for the final evaluation. This evaluation was different from the usability test in a number of ways: (i) the evaluation was conducted in a real environment, at a place with an active redevelopment process; (ii) the survey in the prototype was more carefully designed with the help of experts; (iii) the participants volunteered because of their interest in the concept; and (iv) the participants just completed the survey, without performing predefined tasks. These factors, along with a refined prototype, resulted in valuable feedback on the potential of the concept. To start with, tasks for the survey were defined by architects. Their contribution was a set of questions with possible answers.

When the survey was ready, participants were recruited from the Physical Planning program at Department of Urban Design and Planning, four people volunteered because of their interest in the concept, 2 males and 2 females. Choosing Urban Design and Planning students had a twofold role, on one side they were close to the age of the target user, on the other they had knowledge about planning and could see the concept also from the perspective of an urban planner. After the in-app survey was completed by the users, participants had the opportunity to post suggestions and interact with the rest of the app. A semi-structured group interview was held afterwards.

Results. The contribution from the architects to create a survey worked as an evaluation of task types, and two of the task types in DELTA, multiple choice and check-boxes, were introduced in this phase. Creating surveys emerged as a process requiring a good degree of situatedness. For example, when the defined questions were mapped to specific places, some of the questions were far away from any other question, and it would be difficult to describe the path between them without using pins on the map. To solve this problem, some of the more general questions not referring to a specific place were placed strategically to create a route without very long distances between each task. Also, one task had to be removed, since the whole area was closed by a fence.

During the evaluation, the participants provided suggestions for improving engagement and effectiveness of the solution. One suggestion was to physically tag urban objects using the smartphone or draw upon pictures of them. It was unclear how this would support public participation, but was perceived as a way to increase the awareness of the public space. One participant declared that she was motivated in reaching the next location by the distance indicator. However, some kind of reward was expected at the end of the treasure hunt. The time spent finding tasks was suggested to be used as a performance indicator to assign extra points to the users. Some users suggested to allow participants to add different symbols on the map in order to provide specific feedback connected to the exact point where the participant is located. The symbols could simply be a green and red mark to indicate positive and negative places. On the overall, the evaluation of DELTA was positive and all participants evaluated it as a tool with good potential to improve participation among young people.

6 Discussion

Results from the evaluation of DELTA confirmed the effectiveness and attractiveness of the concept. Situated action was perceived as useful, especially when the geographic area was not usually frequented. Situated engagement in this context fits especially well, compared to just sitting in a room browsing maps or other support material. A challenge connected to deploying the app for a short time in a certain context, is that it might not give enough time for the reward system to reach its full potential, which requires users to post suggestions and interact with them during time. However, if the participants are sufficiently engaged in the app and the context, they can be willing to continue using the app also when the organized event is over.

During our tests, all the users managed to correctly complete all the tasks without help. From the questionnaire, participants perceived the app as very usable. It is here important to underline that most of the participants interacted more with the environment than with the smartphone when navigating between the tasks. This is a positive outcome compatible with the objective of increasing awareness of the surroundings environment and consequently improving quality in the contribution.

Suggested improvements included adding some activities along the route of the survey, providing some hints about how the next location is like and publishing the results in order to freely share the contributions. These suggestions can be interpreted as the need for more diverse engagement mechanics based on the physical environment and the recognition of a significant outcome, perceived important enough to be shared with the community to facilitate the change.

Participants highlighted the limitations of traditional methods and the increased awareness provided by DELTA: (i) maps can be easily misunderstood; (ii) it's easier to provide feedback when physically located in the context; (iii) moving around allows contributing from a new perspective, increasing the quality of the feedback. The game elements adopted to shape the survey served well for the purpose and were positively accepted during the tests, contributing in engaging participants. Game elements in the prototype were intended to also motivate and support continuous participation. However, this outcome cannot be confirmed without testing over a longer period. Also, it remains unsure whether or not the app would be used by young people on their own initiative, although the youth city council member that we interviewed during the first iteration suggested that people would be willing to engage on their own initiative as long as the projects were also seen as interesting and close to their home.

7 Conclusions

The research presented in this paper investigates how to support participation of young people in urban planning. Based on a review of methods of participation, we decided to focus on situated engagement and include game elements in the design of an application for smartphones. A state of the art analysis on public participation apps was performed to ground the work and build on top of current research on the topic. DELTA was then developed and tested in three iterations. The evaluation of the app in the three phases was very positive and some of the experts have expressed their willingness to try it out to promote participation in some of the controversial redevelopment projects currently ongoing in the city.

Technically, this requires to port the app to other platforms, so that it can be used in a large scale field study. Future work will also focus on how DELTA can be integrated into e.g. school activities so to motivate usage, but also to improve its learning impact. Finally, an in-app or web-based interface for planners needs to be created. Until now we have focused on using the app, rather than creating projects and surveys. This is however critical if the app has to be used on a regular basis. As part of this work it will be interesting also to consider how surveys could be created not only by city planners or teachers, but also by young people themselves. This will allow to move one step further in the ladder of participation (Hart 1992).

References

Allen M, Regenbrecht H, Abbott M (2011) Smart-Phone Augmented Reality for Public Participation in Urban Planning, pp 11–20

Bartlett S (2002) Building Better Cities with Children and Youth

Bohøj M, Borchorst NG, Bødker S, Korn M, Zander, P-O (2011) Public deliberation in municipal planning: supporting action and reflection with mobile technology. In: Proceedings of the 5th international conference on communities and technologies, pp 88–97. ACM

Breitbart MM (1995) Banners for the street: reclaiming space and designing change with urban youth. J Plann Educ Res 15(1):35–49

Castells M, Fernandez-Ardevol M, LinchuanQiu J, Sey A (2009) Mobile Communication and Society: A Global Perspective. MIT Press, Cambridge

Chawla L, Heft H (2002) Children's competence and the ecology of communities: a functional approach to the evaluation of participation. J Environ. Psychol. 22(1–2):201–216

Clark BY, Brudney JL, Jang S-G (2013) Coproduction of government services and the new information technology: investigating the distributional biases. Public Adm Rev 73(5):687–701

Corsi M (2002) The child friendly cities initiative in Italy. Environ Urbanization 14(2):169–179

Cunningham CJ, Jones MA, Dillon R (2003) Children and urban regional planning: participation in the public consultation process through story writing. Children's Geographies 1(2):201–221

Dennis SF (2006) Prospects for qualitative GIS at the intersection of youth development and participatory urban planning. Environ Plan A 38(11):2039–2054

Francis M (1988) Negotiating between children and adult design values in open space projects. Des Stud 9(2):67–75

Frank KI (2006) The potential of youth participation in planning. J Plan Lit 20(4):351–371

Gleeson B, Sipe N (2006) Creating Child Friendly Cities: New Perspectives and Prospects. Routledge

Gordon E, Schirra S, Hollander J (2011) Immersive planning: a conceptual model for designing public participation with new technologies. Environ Plan B Plan Des 38(3):505–519

Haider J (2007) Inclusive design: planning public urban spaces for children. In: Proceedings of the institution of civil engineers-municipal engineer, pp 160:83–88. Thomas Telford Ltd

Halprin L, Hester Jr RT, Mullen D (1999) Lawrence Halprin [Interview]. Places 12 (2)

Hart, RA (1992) Children's Participation: From Tokenism to Citizenship, vol 4. Innocenti Essays. ERIC

Hirsch AB (2011) Scoring the participatory city: lawrence (& Anna) halprin's take part process. J Architectural Educ 64(2):127–140

Horelli L (1997) A methodological approach to children's participation in urban planning. Scand Hous Plan Res 14(3):105–115

Innes JE, Booher DE (2004) Reframing public participation: strategies for the 21st century. Plan Theory Pract 5(4):419–436

Irvin RA, Stansbury J (2004) Citizen participation in decision making: is it worth the effort? Public Adm Rev 64(1):55–65

Kangas A, Rasinmäki J, Eyvindson K, Chambers P (2015) A mobile phone application for the collection of opinion data for forest planning purposes. Environ Manage 55(4):961–971. doi: 10.1007/s00267-014-0438-0

Keele S (2007) Guidelines for performing systematic literature reviews in software engineering. In: Technical Report, Ver. 2.3, EBSE Technical Report. EBSE.sn

Korn M (2013) Situating Engagement: Ubiquitous Infrastructures for in-Situ Civic Engagement. Ph.D. thesis, Aarhus UniversitetAarhus University, Science and TechnologyScience and Technology, Institut for Datalogi Department of Computer Science

Lansdown, G (2010) The realisation of children's participation rights: critical reflections. In: A handbook of children and young people's participation: perspectives from theory and practice

Laughlin DL, Johnson LC (2011) Defining and exploring public space: perspectives of young people from regent park, Toronto. Children's Geographies 9(3–4):439–456

Lehner U, Reitberger, W, Baldauf M, Fröhlich P, Eranti V (2014) Civic engagement meets pervasive gaming: towards long-term mobile participation. pp 1483–1488. doi: 10.1145/2559206.2581270

Mallan K, Greenaway R (2011) Radiant with possibility: involving young people in creating a vision for the future of their community. Futures 43(4):374–386

Malone K (2013) The future lies in our hands': children as researchers and environmental change agents in designing a child-friendly neighbourhood. Local Environ 18(3):372–395

Norwegian Government (2008) Planning and Building Act, June

O'Hara, K (2008) Understanding geocaching practices and motivations. In: Proceedings of the SIGCHI conference on human factors in computing systems, pp 1177–1186. ACM

Oksman V, Väätänen A, Ylikauppila M (2014) Co-creation of sustainable smart cities: users, participation and service design. pp 189–95

Passon C, Levi D, del Rio V (2008) Implications of adolescents' perceptions and values for planning and design. J Plan Educ Res

Reiersølmoen M (2016) Facilitating children and youth's participation in urban planning

Roberts N (2004) Public deliberation in an age of direct citizen participation. Am Rev Public Adm 34(4):315–353

Schröder C (2014) A mobile app for citizen participation. November 2014, pp 75–78. doi: 10.1145/2729104.2729137

Spier, JJ (2013) A walk in the park: an experiential approach to youth participation. Youth Studies Australia 32 (3)

Torres J (2012) Participation as a pedagogy of complexity: lessons from two design projects with children. Urban Des Int 17(1):62–75

Tullis TS, Jacqueline NS (2004) A comparison of questionnaires for assessing website usability. In: Usability Professional Association Conference, pp 1–12

Wilks J, Rudner J (2013) A Voice for Children and Young People in the City. Aust J Environ Educ 29(1):1–17

Augmented Reality and Mobile Learning in a Smart Urban Park: Pupils' Perceptions of the EduPARK Game

Lúcia Pombo[✉], Margarida Morais Marques, Vânia Carlos, Cecília Guerra, Margarida Lucas, and Maria João Loureiro

Research Centre on Didactics and Technology in the Education of Trainers - CIDTFF, Laboratory of Digital Contents, Laboratory for Supervision and Evaluation, University of Aveiro, Aveiro, Portugal
{lpombo,marg.marq,vania.carlos,cguerra,mlucas,mjoao}@ua.pt

Abstract. The EduPARK game is developed under a game-based learning methodology. It is designed for outdoor learning settings by employing geocaching principles and mobile Augmented Reality technologies. The game aims to develop users' authentic and autonomous learning about diverse interdisciplinary themes in a smart urban park. It integrates learning guides for different target groups of basic education. The purpose of this paper is to present the game prototype development, which followed a design-based research approach. The evaluation of the game involved 74 pupils from two school levels (aged 9–10 and 13–14). They explored the game and their reactions were registered. Focus groups were conducted at the end of the experience. The evaluation allowed identifying positive characteristics of the game, such as immediate feedback and collaborative dynamics. Some questions included in the learning guides were perceived as difficult to understand and also some features came out to be considered for future improvements.

Keywords: Augmented reality · Mobile learning · Smart urban park · Educational games

1 Introduction

As pupils' access to mobile devices, such as laptops, tablets, smartphones and video game consoles, increases in several contexts, the debate around mobile learning (Clarke and Svanaes 2015) and its educational potential becomes more critical. The ubiquity of mobile devices extends learning, both in formal and informal settings, and when combined with Augmented Reality (AR), it has the potential to move learning to outdoor settings.

AR is a technology that enhances life experiences by employing virtual elements in real time (Dunleavy 2014). It enables pupils to be placed at the center of ubiquitous educational contexts and to interact with digital information embedded into physical environments (Gianni and Divitini 2015). In a recent report, authors point out that AR amplifies access to and interaction with information, hence, creating new learning opportunities for broader understandings (Johnson et al. 2016). Several other studies

Ó. Mealha et al. (eds.), *Citizen, Territory and Technologies: Smart Learning Contexts and Practices*, Smart Innovation, Systems and Technologies 80, DOI 10.1007/978-3-319-61322-2_9

(Radu 2012; Pérez-Sanagustín et al. 2014; Akçayır and Akçayır 2017) suggest that AR enhances pupils' enjoyment, motivation and interest to learn. For example, Akçayır and Akçayır (2017) highlight that this type of technology provides immediate feedback and supports autonomous learning, which can have a positive effect on pupils' motivation and increase their learning performance. Moreover, AR has been shown to be able to reduce cognitive load through the annotation of real world objects and environments as well as to increase long-term memory retention (Santos et al. 2014). However, for such affordances to occur, the multimedia material should have curricular and educational relevance (Radu 2014).

AR supported by mobile devices can move learning to outdoor settings, such as Smart Cities (SC). This concept is closely related to using *smart* technology to improve city life. Studies in SC as a context for learning (smart education) show the potential of the adoption of mobile technologies to generate and collect data for situated games in the city (Gianni and Divitini 2015), namely in the so called Smart Urban Parks (SUP). SUP are based on mobile learning, i.e. on anywhere and anytime personalized learning (Naismith et al. 2004). They foster authentic and situated learning outside the classroom (Jonassen 1994), but also personal and collaborative learning within a lifelong perspective (Naismith et al. 2004). SUP are also considered contexts that can be used to promote new modes of learning in science education, for instance concerning environmental education, since the ability to understand ecosystems is enhanced by experiences in real environments (Kamarainen et al. 2013). Moreover, they have the potential to provide learning experiences that value biodiversity (Ballantyne and Packer 2002), and attract not only pupils and teachers, but also a wide range of tourists and local visitors (Ballantyne et al. 2008), especially if associated with the use of AR and mobile technologies.

One of the emerging potentials of mobile technologies exploration in educational contexts is related with digital games (Prensky 2007). Future developments in this area involve evaluating and analyzing game usage data, providing powerful tools on how to create better learning experiences, and developing game-based learning, supported by significant data about the pupils' perception and their performance while playing (Groff et al. 2015). Additionally, the competition created by games may increase pupils' engagement in challenging learning situations and improve their overall sense of enjoyment. When game's wining conditions require working with other players, collaborative dynamics can also be promoted (Hwang et al. 2015).

The EduPARK project aims to contribute to the SUP concept by designing, implementing and evaluating the EduPARK game, supported by a mobile app, to promote learning within the urban park *Infante D. Pedro*, located in Aveiro (Portugal). This game includes several learning guides for different target groups (pupils, teachers and, possibly, tourists) and is supported by geocaching principles (hunting treasures/caches with the support of technology). The innovation of this project relies on the articulation of (i) new and easy to explore technologies; (ii) geocaching games; and (iii) multidisciplinary educational resources. The beta version of the EduPARK game was tested in the above-mentioned SUP in order to gather pupils' perceptions of the game as a means to improve it. The project methodology follows a design-based research approach and this work reports the implementation and evaluation phase (Parker 2011) of the first cycle.

In the following sections, we briefly describe the EduPARK game, the methodological options, including the data gathering and analysis processes, as well as the results and their discussion. In the final remarks section, empirical-based recommendations are proposed for the improvement of the EduPARK game and for future work.

2   Development of the EduPARK Game

The EduPARK project proposes an activity that combines AR and geocaching games in a SUP, supported by a mobile app. At the present stage, a beta version of the app was already conceived. This first version comprised an interactive AR quiz-based game to be played by teams of three or four pupils, in a friendly competition approach. Each team needs to be accompanied by one adult monitor for safety reasons and also to collect observation data. The basic structure of the app is summarized in Fig. 1.

One of the initial screens of the app prompts the players to identify their team and select a learning guide (**a** in Fig. 1): one for First Cycle pupils (aged 9–10) and another for Third Cycle ones (aged 13–14). Only learning guides for these two Cycles were included in the app, because its beta version was to be tested by a convenience sample of pupils as explained in methodological options section.

The quiz questions, as well as the predefined path in the SUP, are different depending on the selected guide. A short tutorial (**b** in Fig. 1) explains how to use the camera tool to recognize the AR markers. These unlock the access to information relevant to answer a series of questions related to each specific location. Next, the players can initiate the cycles (example in **c** in Fig. 1): (i) following instructions to find a specific AR marker; (ii) using the device to recognize the prompted marker; (iii) accessing a set of multiple-answer questions; and (iv) receiving adequate feedback to answers and scores, if answered correctly. The app also provides feedback through the constant display of accumulated scores and offers a sense of progress through the number of questions answered, locations visited and caches discovered vs. the total number of these items. The app integrated the search for and discovery of three physical caches in the SUP.

To support the players' progress, the app provides a number of tools: camera (to recognize AR markers and take pictures), backpack (to see the pictures taken), compass (to support the players' orientation in the park) and a map of the park (with the players' location as well as the next location or cache to visit). At any time, the players can access the *help menu*, accessed through the blue button available at the top of the majority of screens. This menu has a *general help* screen explaining the meaning of the symbols used in the general screen of the game (see **d** in Fig. 1), and a *help screen* for each tool of the game. Finally, the last screen (see **e** in Fig. 1) displays the overall performance of the team, with the total score, the number of correct and wrong answers and the completion time of the game.

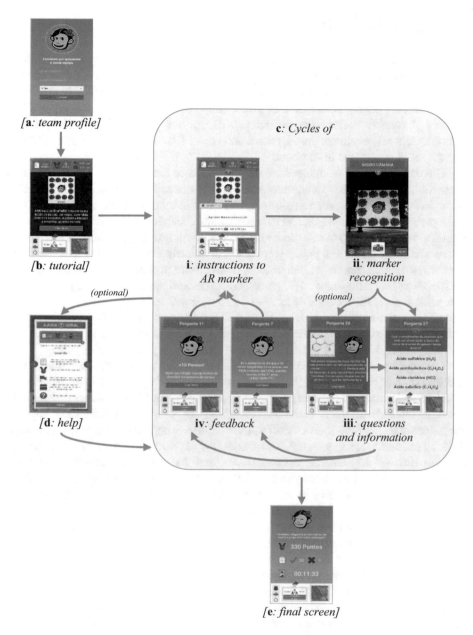

Fig. 1. Structure of the EduPARK app, illustrated by print screens of the beta version of the app, available only in Portuguese language

3   Methodological Options

As the focus of the EduPARK project is the development of a learning intervention in a real educational context, involving multiple iterations for refinement and evolution of a mobile AR game, a design-based research approach was considered suitable to achieve its objectives (Anderson and Shattuck 2012). This approach includes two or more cycles of four phases: 1. analyze the problem; 2. design and develop potential solutions; 3. implement and evaluate; and 4. reflect and report (Parker 2011). The present paper reports the results of the third phase.

The data gathering techniques selected to evaluate the game were an observation protocol and a focus group interview at the end of the activity, allowing triangulation (Amado 2014). The observation protocol was filled in by monitors and comprised two main parts: (i) a think aloud section to register pupils' behavior and perceptions; and (ii) a critical incidents section to collect information about problematic or positive events. At the end of each session, pupils were randomly distributed in two groups in order to conduct two simultaneous but independent focus groups, to facilitate the sharing of their perspectives about the EduPARK game and associated app. Focus groups have been recognized as useful tools for pilot tests in educational research, as they allow inter-viewees to explain their experience in depth (Williams and Katz 2001). All pupils of each focus group belonged to the same class, hence, were familiar with each other. Two focus groups had 11 pupils (Third Cycle) and four had 13 pupils (First Cycle).

The pupils were asked to: (i) classify (on a 1 to 5 scale, being 5 the maximum score) and justify their enjoyment of the experience, so that perceptions of their satisfaction could be understood; (ii) classify (using the same scale) and justify the easiness of the game, so that difficult features could be identified; (iii) propose suggestions to improve the game; and (iv) give their overall opinion of the experience. The interviews were audiotaped, and were moderated with flexibility, varying from 8 to 21 min, with an average time of 15 min. Observation notes and focus groups transcriptions were submitted to content analysis (Amado 2014), aiming to uncover the game positive features and the ones needing improvement. The categories emerged from the empirical data and are described in the next section.

The implementation and evaluation of the EduPARK game were conducted under the Open Week of Science and Technology of the University of Aveiro, in November 2016. The main purpose was to gather pupils' perceptions to improve the EduPARK game. This evaluation focuses on the pedagogical features of the EduPARK game. The technical evaluation of the app is described in another work (Pombo et al. in press). During the inscriptions period to the evaluation activities, two teachers of the First Cycle and one of the Third Cycle showed their strong interest in participating. Hence, the convenience sample of pupils/users of the app. Their characterization is showed below (Table 1).

At the beginning of the game, each group received a mobile device with the app, and the monitors presented the game and explained how to use the app.

The next section presents the results of the preliminary evaluation of the EduPARK game prototype and discusses them accordingly to the literature reviewed whenever possible. The anecdotal evidence provided improvement suggestions that will be presented in Sect. 4.5.

Table 1. Characterization of the classes involved in the evaluation of the EduPARK game

Group	Cycle	N. of pupils	Average age	% of female	% of male
1	1st	26	9.0	69.2	30.8
2	1st	26	9.0	50.0	50.0
3	3rd	22	12.9	31.8	68.2

4 Results and Discussion

Data collection gathered from participants (focus groups) and from observations was both broad and specific, leading to concrete suggestions for improving the experience of using the EduPARK game in loco. Positive and negative perceptions of the EduPARK game are presented by categories, namely regarding enjoyment and level of difficulty. Content analysis also allowed identifying improvement suggestions for the development of new versions of the game.

4.1 Positive Perceptions of Enjoyment

First, pupils were asked to classify the activity using a scale, in which 1 stood for lower enjoyment and 5 for higher enjoyment. The answers revealed that in all focus groups, except one, the classification was 5. This implies that the activity was well rated by the pupils. This result is in line with studies mentioned before that point out that AR tools promote pupils enjoyment (Radu 2012; Pérez-Sanagustín et al. 2014; Akçayır and Akçayır 2017). Justifications provided by the pupils were diverse, ranging from perceptions that could imply the motivational value of the activity to the valorization of the outdoor activities. In the following paragraphs we describe pupils' justifications, illustrating them with examples.

The first subcategory is related with 'increased motivation', as illustrated by the citation: "The activity was enriching because it helped us to develop values and helped us to wish for more learning ..." (G3 pupil) and "Pupils said this activity is really fun and cool" (G2, Observer H). This result is in line with the literature that reports that AR and digital games can promote motivation (Kamarainen et al. 2013; Pérez-Sanagustín et al. 2014; Dunleavy 2014; Johnson et al. 2016).

Pupils valued several aspects of the activity. Among them is the 'valorization of the social aspect of the activity' as mentioned by two pupils: "I liked it because we are socializing with our friends" (G1 pupil). Those results are similar to those of (Bacca et al. 2014) that reports socialization with peers as one of the advantages of AR technology. Another aspect was related with 'valorization of the outdoor activity': "I think we can achieve better results outside the classroom, because we are in physical and visual contact with the content we are supposed to learn" (G3 pupil). The possibility to establish connections with content was also reported by (Bacca et al. 2014) and it was acknowledged to support situated and authentic learning (Ballantyne and Packer 2002; Naismith et al. 2004). Pupils also pointed to the 'valorization of the learning pace', as the citation shows: "… we learn quicker" (G3 pupil), and to the 'valorization of the immediate feedback' on the correct and incorrect answers included in the EduPARK game. For instance, one pupil stated: "I enjoyed it, because if we answered wrongly, the correct answer would show and we could learn more" (G2 pupil). The immediate feedback is related with increased learning performance (Kamarainen et al. 2013). This feature provides an individualized learning strategy to heterogeneous groups of pupils, giving an extra scientific explanation of the learning content activities integrated in the interactive quiz-based game. This is also one of the reported advantages of AR technologies in the literature, one that can promote autonomy (Kamarainen et al. 2013).

4.2 Negative Perceptions of Enjoyment

As described above, the majority of the pupils pointed out positive features of the activity, but they did not provide negative justifications. Nevertheless, some pupils highlighted some negative aspects of the activity related with the level of difficulty, which are presented below.

4.3 Positive Perceptions of the Level of Difficulty

Concerning the level of difficulty, the pupils' perceptions were rated between 4 and 5. Two groups did not justify their classification. The ones who did provided the following justifications.

'Connection with the curricular content' was one reason that pupils pointed out for considering the activity easy, as illustrated by the citations: "As we already knew the content, it was easier" (G1 pupil) or "Pupils stated that they already knew the information about the European holly [*Ilex aquifolium*]" (G2, Observer F). This result is in line with some authors' recommendation concerning AR activities that they should be educationally relevant for pupils (Akçayır and Akçayır 2017) and contextualized, which seems to be the case of the EduPARK game.

'Problem solving strategies' were observed by several monitors that stated, for instance, that: "[Initially] pupils needed the monitors' help, but then they became more autonomous in solving problems" (G2, Observer G).

'Instruction adequacy' was also observed by the monitors, who mentioned: "Pupils easily understood when they had to move to another location" (G3, Observer I).

4.4 Negative Perceptions of the Level of Difficulty

Pupils justified their lower rates concerning the level of difficulty of the EduPARK game referring to specific challenging aspects. One of the aspects is related to 'difficulties with the vocabulary', especially observed in the younger groups: "Pupils didn't know the meaning of 'fertilizer' and 'honoring'" (G2, Observer F) or "Pupils didn't know what plans of symmetry were" (G1, Observer E).

Concerning the 'location of the AR markers' pupils apparently had different views. Some considered them too easy to find: "I believe they should be physically better hidden [referring to the markers]" (G3 pupil); others too difficult: "The last one [marker] was really hard to find" (G3 pupil). Geocaching aspects of the game were also pointed out by the monitors. For example: "Pupils didn't understand how to use the compass" (G2, Observer F) or "Pupils didn't find the right direction" (G1, Observer E).

4.5 Improvement Suggestions

Pupils' improvement suggestions emerged during the activity (registered by the monitors) and the focus groups. They were related with three subcategories: (i) dynamic of the activity, (ii) types of questions, and (iii) interest of the activity.

Concerning the dynamic of the activity, several subcategories emerged. For example, pupils' opinions about the 'teams' constitution' were not consensual, as some of them preferred to work in smaller groups: "I think it would be better to play in teams of only two or three pupils" (G1 pupil), and others favored bigger teams, since "Maybe playing in bigger teams, because [more elements] can think better" (G3 pupil). These contrasting opinions can be related to differences between pupils' ages (9–10 and 13–14). Nevertheless, one of the pupils' concerns was related with the collaboration level within the team, which may be created in gaming situations, as claimed by Groff et al. (2015).

The youngest pupils proposed to 'extend the activity': "I think the activity should have more questions and cover more places in the park" (G1 pupil). This fact may be associated with a stronger level of enjoyment with the activity reported by the youngest pupils (who classified the activity with 5 points).

The following subcategories are related with pupils' suggestions for designing other types of questions for the game. Pupils proposed to include 'more subjects' in the learning guide, such as Portuguese, English, Astronomy, and Sports, as well as to include more 'diverse questions'. For instance, one pupil suggested a new type of question, based on visual recognition: "I would like to see questions that ask me to go to a location represented in a photo" (G1 pupil).

Pupils provided valuable hints to increase the interest of the activity related to the inclusion of 'different paths and sites'. For example, pupils proposed: "… we should have more locations. For instance, I think that we could focus more in the lake, since we have a very beautiful lake [in the park]" (G3 pupil); "Different paths should be implemented" (G1, Observer B). Another pupil's suggestion was related with 'preventing cheating behavior', as expressed in the following citations: "I think that the hints should be different from team to team because, when a team is behind, they can copy what the others are doing" (G1 pupil) and "Pupils think that the teams should have staggered starts during the activity" (G1, Observer D).

Finally, pupils also suggested to 'increase the competition', as revealed by these citations: "We could take a photo nearby the caches and then, the best photo would be the winner" (G2 pupil) and "One of the criterions [to win] should be the time, to increase the competition" (G3 pupil).

5   Final Remarks

The development of the EduPARK AR game follows a design-based research approach. In this work we present the implementation and evaluation phase of the first cycle. The game was experienced by pupils in a SUP, the *Infante D. Pedro* park in Aveiro (Portugal). Data gathering techniques included focus groups (with pupils) and observation (made by monitors). The authors acknowledge some limitations, such as the loss of participants' nonverbal cues (Parker 2011), as the interviews were audiotaped and not videotaped. Another aspect to consider is the fact that the pupils were interviewed in a group, which has the potential to standardize the participants' opinions (Parker 2011). However, given the available resources and setting for the conduction of the interviews, these limitations may not affect the results, taking into account that the aim of this work is to collect the players' opinions regarding the activity in loco.

Results suggest that pupils' considered the game enjoyable and easy to play. However, some negative perceptions were also pointed out. These results allowed us to propose the following design principles for educational games for SUP. The game activities should:

- increase pupils' motivation to learn (Pérez-Sanagustín et al. 2014; Dunleavy 2014; Johnson et al. 2016), by providing immediate feedback (Kamarainen et al. 2013) and promoting the socialization among peers (Bacca et al. 2014);
- value the outdoor aspects of the activity, as well as the SUP related content that promotes situated and authentic learning (Ballantyne and Packer 2002; Naismith et al. 2004);
- allow contact with nature, which seems to promote learning at a faster pace than in the classroom and may increase learning performance (Kamarainen et al. 2013);
- offer opportunities to learn local culture and history issues;
- be connected with the curricular content (Bacca et al. 2014) and employing problem solving strategies in order to develop autonomous learning;
- provide adequate instructions, by attending to eventual difficulties to interpret the game questions and using suitable vocabulary. If support is given, new vocabulary can be introduced;
- be challenging, for instance, by balancing the difficulty of the AR markers localization.

The above-mentioned design principles may contribute to create better learning experiences supported by significant data retrieved from pupils' perceptions and their performance while playing (Groff et al. 2015). Pupils' offered several relevant improvement suggestions, such as: increase the activity length, provide different paths and sites in the SUP, increase competition to promote enjoyment and learning (Prensky 2007) while enabling collaboration (Groff et al. 2015), as well as diversify the type of questions and of disciplines involved. These suggestions will be considered in future work under the EduPARK project. The results show that combining mobile technology with outdoor gaming activities allows learning to move beyond traditional classroom environments that pupils can explore and, simultaneously, make connections with curricular content. Furthermore, the EduPARK game provides collaborative, situated and authentic learning. It also offers new challenges, opens up horizons and opportunities for Science Education. The EduPARK game already integrates some of these recommendations [dynamic of the activity, types of questions, and interest of the activity], because the EduPARK researchers recognize that the game competition is an important aspect for promoting enjoyment and learning (Prensky 2007). In line with (Kamarainen et al. 2013), it is also acknowledged that the Aveiro SUP has important educational potential to develop formal and informal learning about ecological conservation, biodiversity and city historical patrimony, which will be reinforced in future versions of the EduPARK game.

Acknowledgements. This work was financed by FEDER - Fundo Europeu de Desenvolvimento Regional funds through the COMPETE 2020 - Operacional Programme for Competitiveness and Internationalisation (POCI), and by Portuguese funds through FCT - Fundação para a Ciência e a Tecnologia in the framework of the project POCI-01-0145-FEDER-016542. The authors would also like to thank the EduPARK researchers, the participant pupils, teachers, and monitors.

References

Akçayır M, Akçayır G (2017) Advantages and challenges associated with augmented reality for education: A systematic review of the literature. Educ Res Rev 20:1–11. doi:10.1016/j.edurev. 2016.11.002

Amado J (2014) Manual de investigação qualitativa em educação. Imprensa da Universidade de Coimbra, Coimbra

Anderson T, Shattuck J (2012) Design-based research: a decade of progress in education research? Educ Res 41:16–25. doi:10.3102/0013189X11428813

Bacca J, Baldiris S, Fabregat R, Graf S (2014) Augmented reality trends in education: a systematic review of research and applications. Educ Technol Soc 17:133–149

Ballantyne R, Packer J (2002) Nature-based Excursions: School Students' Perceptions of Learning in Natural Environments. Int Res Geogr Environ Educ 11:218–236. doi:10.1080/103820 40208667488

Ballantyne R, Packer J, Hughes K (2008) Environmental awareness, interests and motives of botanic gardens visitors: Implications for interpretive practice. Tour Manag 29:439–444. doi: 10.1016/j.tourman.2007.05.006

Clarke B, Svanaes S (2015) Updated review of the global use of mobile technology in education, London

Dunleavy M (2014) Design principles for augmented reality learning. TechTrends 58:28–34. doi: 10.1007/s11528-013-0717-2

Gianni F, Divitini M (2015) Technology-enhanced smart city learning: a systematic mapping of the literature. Interact Des Archit J, 28–43

Groff J, Clarke-Midura J, Owen VE et al (2015) Better learning in games: a balanced design lens for a new generation of learning games. MIT Education Arcade, Cambridge

Hwang G-J, Wu P-H, Chen C-C, Tu N-T (2015) Effects of an augmented reality-based educational game on students' learning achievements and attitudes in real-world observations. Interact Learn Environ 4820:1–12. doi:10.1080/10494820.2015.1057747

Johnson L, Becker SA, Cummins M et al (2016) The NMC Horizon Report: 2016 Higher, Education edn. Austin, Texas

Jonassen DH (1994) Thinking technology: toward a constructivist design model. Educ Technol 34:34–37

Kamarainen AM, Metcalf S, Grotzer T et al (2013) EcoMOBILE: Integrating augmented reality and probeware with environmental education field trips. Comput Educ 68:545–556. doi: 10.1016/j.compedu.2013.02.018

Naismith L, Lonsdale P, Vavoula G, Sharples M (2004) Literature review in mobile technologies and learning. Bristol, UK

Parker J (2011) A design-based research approach for creating effective online higher education courses. In: 26th Annual research forum: educational possibilities. Western Australian Institute for Educational Research Inc., Fremantle

Pérez-Sanagustín M, Hernández-Leo D, Santos P et al (2014) Augmenting reality and formality of informal and non-formal settings to enhance blended learning. IEEE Trans Learn Technol 7:118–131. doi:10.1109/TLT.2014.2312719

Pombo L, Marques MM, Afonso L, et al (in press) An experience to evaluate an augmented reality mobile application as an outdoor learning tool

Prensky M (2007) Digital game-based learning, 2nd edn. Paragon House, St. Paul

Radu I (2014) Augmented reality in education: A meta-review and cross-media analysis. Pers Ubiquit Comput 18:1533–1543. doi:10.1007/s00779-013-0747-y

Radu I (2012) Why should my students use AR? A comparative review of the educational impacts of augmented-reality. In: ISMAR 2012 - 11th IEEE international symposium on mix augment reality 2012, Science Technology Paper, pp 313–314. doi:10.1109/ISMAR.2012.6402590

Santos M, Chen A, Taketomi T (2014) Augmented reality learning experiences: Survey of prototype design and evaluation. IEEE Trans 7:38–56

Williams A, Katz L (2001) The use of focus group methodology in education: Some theoretical and practical considerations

Becoming a Mobile Internet User in a South African Rural Area: The Case of Women in Dwesa

Lorenzo Dalvit[✉] and Mfundiso Miya

Rhodes University, Grahamstown, South Africa
l.dalvit@ru.ac.za, mfundisomiya@gmail.com

Abstract. The pervasiveness of mobile phones enables people in marginalised contexts, including African rural areas, to access the Internet. Evidence suggests that women are increasingly at the forefront of ICT adoption. This paper explores how and why women in Dwesa, a South African rural area, learn to access the Internet on their mobile phones. Extensive research in this area was analysed to provide a solid background to a small-scale, in-depth qualitative study. Our findings reveal that rural women can be digitally literate and information-aware users who are deeply embedded in local social networks and use their phone to mediate the local context. This is significant as it contradicts a stereotypical image of African rural women as marginal participants in the information age.

1 Background

Mobile phones have been diffusing faster than any other technology in history, as observed by various scholars and agencies (Castells et al. 2007; GSMA 2015; Goggin 2006). Ahonen (2008) notes there are "twice as many cell phones as television sets, three times as many cell phone subscribers as internet users, four times as many cell phones as personal computers, five times as many cell phones as cars". In fact, half (3.6 billion) of the world's population now has a mobile subscription and 60% of these are smartphones (GSMA 2015). Mobile phones have become the technology of choice for developing countries to reduce the connectivity gap (Castells et al. 2007). Aker and Mbiti (2010) note that the adoption of mobile phones in some African countries have exceeded expectations. Africa is the second largest mobile phone market in the world after Asia. Mobile phone penetration in Africa is 67% (Adepetun 2015) and according to Frost and Sullivan (2015), it is expected to rise up to 79% in the next 5 years.

Mobile phones play a major role in internet access in the developing world. In Africa, most people access the internet for the first time on their mobile phones (Donner 2008). The emergence of mobile internet coupled with the widespread uptake of internet-enabled mobile phones contributes to Internet adoption (Deen-Swarray 2016). Since 2010 mobile is the primary form of internet access in South Africa exceeding dial-up connections (Goldstuck 2010). As of the end of 2014, the mobile penetration rate in South Africa was 133%, which means many people use more than one device (Fripp 2014). As reported by the KCPB Internet Trends 2015 47% of mobile phones in South Africa are smartphones (My Broadband 2015). Mobile phones have not only changed

© Springer International Publishing AG 2018
Ó. Mealha et al. (eds.), *Citizen, Territory and Technologies: Smart Learning Contexts and Practices*,
Smart Innovation, Systems and Technologies 80, DOI 10.1007/978-3-319-61322-2_10

the way in which people communicate but also the way they consume media and interact. The mobile phone has been able to merge all previous media into one device. Moore (2007) states, "the consumption of news, the playing of music, watching TV, listening to radio, even viewing movies are all possible on a mobile device. And the internet's two unique capabilities, interactivity and search, are also available on the mobile platform" (Moore 2007). Internet-enabled mobile phones, i.e. devices that can access the internet or successfully run an internet application (Ezemenaka 2013), are accessible even in remote rural areas (see Dalvit et al. 2014; Goldstuck 2013; Collopen 2015).

Obijiofor (2015) found evidence that people in rural areas are using mobile phones to improve their living conditions i.e. looking for employment opportunities, knowledge acquisition, cultural preservation and the promotion of minority languages. Research into mobile phones has concentrated in urban areas (Clayton-Powell 2012; Kreutzer 2009; Kalba 2008). Although over 60% of South Africans reside in urban areas, most people in poorer provinces such as the Eastern Cape live in rural areas (Statistics South Africa 2011; Statistics South Africa 2015). People in rural areas mainly access the internet on their mobiles (RIA 2012). Over 90% of people in rural areas use prepaid mobile services due to lower or irregular incomes and the fact that prepaid services do not require a bank account or proof of physical address (Esselaar and Weeks 2007). As prepaid generally costs more than contracts (e.g. the cost per megabyte is higher) mobile communication is comparatively more expensive for rural than for urban dwellers. Pade-Khene et al. (2010) note rural households in rural Eastern Cape spend an average of R160 out of an income of R1000 every month on mobile communication. This represents a substantial portion of disposable income. A comparative study of township and rural students shows that, despite similar levels of access to mobile phones and use for communication, rural students lagged behind by between 22% and 37% in internet-related activities i.e. web browsing, email, social networks etc. (Gunzo and Dalvit 2012). A study conducted in a rural area in the Eastern Cape shows 50% of young people perform networked related activities when compared to 15% of older people (Dalvit and Strelitz 2013). Women perform more networked activities and money-related activities than men and the gender gap increases with age. Qualitative follow up suggests that (particularly older) women have access to phones with more advanced features because they are the recipients of social grants (Dalvit and Strelitz 2013).

Women appear to be at the forefront of ICT adoption in Dwesa, a rural area on the Wild Coast of the former homeland of Transkei (Mapi et al. 2008). For over a decade, the area has been the site of the Siyakhula Living Lab, an ICT-for-development project, and human and social dynamics related to technology in the area are well documented (Thinyane et al. 2008; Pade-Khene et al. 2010; Christoferi 2015). A multidisciplinary group of researchers from Rhodes University and the University of Fort Hare engaged in extensive digital literacy training ranging from computer literacy to the use of mobile phones as creative tools (see Dalvit et al. 2006; Dalvit 2015a). In both research and training activities, women constitute an overwhelming majority of the participants. Villages adjacent to the Dwesa nature reserve are equipped with networked computer labs, which serve as points of access to the internet for the neighboring communities. A study conducted by Gunzo and Dalvit (2014) suggests that 70% of young people in the area have access to a mobile phone and that mobile phones are the most common ICT devices used in the community. Moreover, they noted a 20% yearly increase in the proportion of users who have access to an internet enabled mobile phone. Since many people have access to a mobile phone in Dwesa, a substantial amount of time and money is invested on the internet on these mobiles. Networked activities (internet-related activities such as web browsing, email, social networks) are performed by most people i.e. 57% are using instant messaging and 64% are on social networks (Collopen 2015).

The present study focuses on a well-researched site to complement existing knowledge in terms of: (1) the reasons and processes of acquisition of mobile devices as well as relevant skills; (2) the specific conditions and challenges of mobile use in rural areas; (3) the perspective of women. Exploring how and why women in Dwesa become mobile internet users could inform research and policy efforts focusing on the role of technology in rural areas. Apart from a head start in digital literacy training and access to ICT in schools, Dwesa is representative of many rural contexts in the country with a lack of basic services, lack of infrastructure, high unemployment rates. As many other rural areas (see Goliama 2011) Dwesa is a very traditional and largely patriarchal society. Due to urban migration of working-age men, women constitute the backbone of the local economic and social life (Pade-Khene et al. 2010). In 2005, the United Nations recognised that ICT could be used as a tool for women empowerment in the developing world. Through effective use and access, ICT can improve literacy levels among women, provide employment and entrepreneurship opportunities and promote gender equality (UN 2005). Buskens and Webb (2009) note that although ICT policies are being implemented in Africa there is a lack of knowledge about how gender inequalities and ICT affect each other, and it becomes important to understand how ICT can play a role in improving the lives of women in developing countries. Hilbert (2011) notes that compared to previous ICT, mobile phones show greater potential to bridge the gender digital divide and empower women (see Dalvit 2015b and Buthelezi 2015 for some examples from Dwesa).

2 Theoretical Framework

The diffusion of an innovation such as mobile internet use in a rural area can be understood in terms of characteristics of the innovation, communication channels used, time it takes to diffuse and the relevant social system (Rogers 2003). Members of a social system can be classified according to their role in the adoption process as innovators, early adopters, early majority, late majority and laggards. Holden (2012) emphasizes that technologies are embedded within a system of social structures that determines how they diffuse. In Dwesa, educated and relatively mature women (e.g. the teachers discussed in Mapi et al. 2008) can be considered innovators, i.e. they launched ICT use into the social system and assumed a gate-keeping role. Qualitative research (Buthelezi 2015; Dalvit 2015b) suggests that women play a key role in adopting and conveying internet use through interpersonal networks (early adopters) and providing interconnectedness to the systems interpersonal networks (early majority). Quantitative research (see Gunzo and Dalvit 2014; Collopan 2015) suggests that mobile internet use is reaching the point at which enough individuals use the internet to make further adoption self- sustaining (Rogers 2003). At the individual level Rogers (2003) identifies five steps in the adoption process: knowledge; persuasion; decision; implementation and evaluation. Goldstuck (2010) notes that after gaining access to the Internet, it may take up to 5 years for a person to become an active internet user in South Africa. What this author refers to as the Digital Participation Curve may account for the progression between different steps in the adoption process.

It is important to distinguish between physical and epistemological access to the internet. The former refers to ownership or sharing of a device that can access the internet. The availability of cheap smartphones allowed for rural communities to access the internet (Dalvit 2015b). In some African contexts, young (unmarried) women, may be considered disreputable if they own mobile phones as it is assumed they have received them through illicit romantic relationships (Sey 2011). Mobile phone sharing increases access and may contribute to digital inclusion in a rural context (Dalvit and Strelitz 2013). Mobile phone sharing is a common practice in most African households where mobiles are frequently used as household 'items' by family members (Aker and Mbiti 2010; ITU 2013). In some instances, married women use their husband as proxy to make a call (Burrel 2010). In other cases, however, women may exploit their better control of mobile technology to gain an insight into and exert covert control over traditional affairs, by acting as proxies to their less ICT savvy husband (Dalvit 2015b).

The ability to effectively use ICT by overcoming barriers such as language, literacy, ICT competence is referred to as epistemological access (du Plooy and Zilindile 2014). With specific reference to Dwesa, Thinyane et al. (2008) note how the Western metaphors shaping ICT tools may pose challenges to the acquisition of digital skills. Digital literacy comprises a set of technical, cognitive as well as sociological skills to navigate digital environments (Eshet-Alkalai 2004). Digital literacy refers to the ability to understand, evaluate and integrate information available on the internet (Gilster in Eshet-Alkalai 2004). Eshet-Alkalai (2004) proposes five types of skills that are vital for digital literacy, which are photo-visual literacy; reproduction literacy; information literacy; branching literacy; and socio-emotional literacy. Photo-visual literacy (the art of visual representation) is the ability to decode and understand visual messages such as icons and symbols instead of wording. This aspect is particularly important in a rural African context, where functional literacy (particularly in English) is relatively low. Reproduction literacy (the art of creative recycling of existing materials) is the ability to create meaningful and creative interpretations by integrating existing information, this allows for originality and creativity when reproducing information online. The proliferation of memes online (Bosch 2013) suggests growing competence as far as this aspect is concerned. Branching literacy is "knowledge construction from information that is accessed in a nonlinear manner" (Eshet-Alkalai 2004). Branching cognitive skills allow for improved navigation performance on the web (Lee and Hsu 2002). Information literacy (the art of skepticism) is the ability to evaluate and assess information properly. Socio-emotional literacy allows users to be critical when consuming content online, where online information can easily be published and manipulated. Eshet-Alkalai (2004) explains that cyberspace is a global village with its own unwritten rule where "socially-literate users of the cyberspace know how to avoid "traps" as well as derive benefits from the advantages of digital communication" (Eshet-Alkalai 2004). As a global village with unwritten rules, users must be very critical, analytical and mature with a high degree of information and branching literacy (Eshet-Alkalai 2004). These digital literacy skills may not be unique to digital environments but allow for informed, creative and navigational online participation and they are necessary to use ICT. Besides formal instruction, people become digitally literate through experience. As an example, when people start accessing the Internet on their phone they might be faced with unexpected expenditures and learn to manage data sparingly (Goldstuck 2013).

People become mobile internet users for instrumental as well as symbolic reasons. Getting access to internet-enabled mobile phones allows individuals to reduce communication costs (Aker and Mbiti 2010). Buthelezi (2015) notes how women in Dwesa save money by using their phone to access social grants and manage their savings as part of local stokvels. Evidence suggests that social media provide an increased sense of security for single mothers and unmarried women (see Dalvit 2015b). Roux and Dalvit (2014) note how for rural women mobile phones are a tool to stay connected not only to members of their communities but also the 'outside' world. Rogers (2003) notes that perceived characteristics of an innovation (e.g. relative advantage, compatibility with ones values, complexity, trialability and observable benefits) may influence the pace of its adoption.

3   Methodology

The goal of the present study was to describe how a group of women of Dwesa became mobile internet users. This was achieved by answering the following research questions:

1. How do women in Dwesa access the internet?
2. How do women in Dwesa become digitally literate?
3. Why do women in Dwesa become mobile internet users?

A descriptive case study approach was employed. Yin (1999) explains that "case studies are the preferred strategy when 'how' and 'why' questions are being posed, when the investigator has little control over events, and when the focus is on a contemporary phenomenon within some real-life context". (Yin 1999) Although a substantial body of research on related topics exists for Dwesa and similar rural areas, the adoption and use of internet-enabled phones is a relatively new phenomenon. Yin (1999) states that "The primary purpose of a case study is to understand something that is unique to the case". The prominent role of women as pioneers and champions in the adoption and use of ICT (Mapi et al. 2008) and particularly mobile phones (Dalvit 2015b) makes Dwesa an interesting, though not necessarily unique, case for the study of how rural women become online participants.

The three research questions were answered by analysing data collected through different methods. Semi-structured interviews were conducted to find out how women in Dwesa make sense of the process of becoming a mobile internet user (see Appendix A). Five in-depth interviews were conducted with women of working age (18–60 years) who have access to an internet-enabled mobile phone. Interviews yielded significant amounts of information about facts as well as opinions from the individual's perspective

(Hancock and Algozzine 2006). Yin (1999) emphasizes that interviews should appear as 'guided conversations' rather than 'rigid or structured queries'. This was particularly important for participants in Dwesa who may not be used to interviews as a research method. This was also ideal to preserve anonymity and to share their experiences without public scrutiny, as would have been the case in a focus group. Women who are classified at different stages of Roger's adoption categories were identified to reconstruct their mobile phone narratives. In response to the research questions, a timeline was created from when the participant first accessed the internet to when they became digitally literate and mobile internet users. Convenience sampling was used to identify participants from previous contacts made during workshops and training sessions in the area, where additional participants were identified to ensure the inclusion of participants from different age groups, with different socio-economic status, level of education etc. Data from the semi-structured interviews were triangulated with observation and field notes of ICT use in daily activities within the community as well as ICT training workshops.

4   Findings

Research participants represented a wide range of socio-economic and educational backgrounds and varied in age and ICT competence. Interviewees could be briefly described as: (A) a mature teacher who participated in formal ICT training courses; (B) a senior person who is casually employed; (C) an unemployed single mother who took part in informal ICT training; (D) a student from a relatively affluent family with computer access at home; (E) an unemployed young woman with very limited means. Possibly due to the inherent bias due to the research topic, i.e. mobile Internet use, most of them would qualify as either innovators or early adopters. Irrespective of their age, they started using Internet on a mobile phone relatively early in their life, which for younger participants meant while they were still at school. All respondents owned a mobile phone and some of them owned two, both Internet enabled. Models would range between entry level and middle range (e.g. Nokia Asha). When talking about their mobile phone, they often referred to its Internet capability as a decisive factor in choosing a particular model. Phone ownership was the norm and age determined the mode of acquisition. Younger women receive their phones as gifts i.e. from parents "I requested my mother to buy me a phone if I passed my matric with good grades", whereas the older women bought their own mobiles, "I got a piece job and managed to save enough money to buy my phone". Evidence suggests that sharing of mobile phones as a household item was limited to specific circumstances, linked to the acquisition of digital skills. A young respondent stated that her mother (an ICT champion in the community) was willing to share her phone so that she could learn to search for information for school and cooking recipes online.

Digital literacy levels varied according to different aspects. Observation and interviews supported claims made by all respondents that they possessed good levels of photo-visual and reproduction literacy. All were familiar with icons and symbols pertaining to online communication and mobile apps. Participants of all ages shared multimedia material such as photos and videos. Younger women also created and shared memes. The relevant skills were acquired through interaction with friends and peers at school or at work. One-way instruction only related to computers, with older women learning about icons in a computer literacy course and a younger one being taught how to use computers by her brother. The idea of learning how to use mobile phones in any way other than social interaction was regarded as amusing.

Information was sought online about a wide range of topics, almost invariably through a Google search. There was no evidence of branching literacy or critical information literacy and what was found on the Web was taken at face value. Despite having taken a university course on ICT in Education, even a mature respondent did not problematise the accuracy or veridicity of what she found online. In terms of socio-emotional literacy, however, respondents seemed aware of the dangers present online and were concerned about their privacy and security. A participant stated she is always on her phone looking for jobs and staying current with news. Her online safety is very important to her and she avoids giving her personal information to strangers, as she is aware of fake job adverts and human traffickers. Another mature respondent expressed her concerns about sharing her personal details online. The same level of concern could not be observed among younger respondents, whose use of the Internet was limited to recreation and education or self-improvement.

Reasons for becoming an online participant were primarily instrumental. The younger unemployed women frequently cited reduction of communication costs as a reason for becoming a mobile internet user "I wanted to chat to my sister in Grahamstown as it is cheaper than making calls". Communication with friends and relatives, particularly those in the cities, was cited by all respondents as one of the main reasons for using the Internet. Other young women stated they learnt about 'things they are not taught about at home' like reading beauty blogs, gaining self-confidence, Facebook Diaries and advice about boys. Being up to date with "what is going on" was emphasised with clear reference to the World outside Dwesa. Older educators become internet users to get access to teaching resources for their learners such as Nalibali (a digital media campaign to promote a reading and writing culture amongst children in all South African languages). Other services only accessible outside Dwesa, such as banking and job seeking were also provided as reasons.

5   Discussion and Conclusions

Being online is important for women in Dwesa and Internet capability was a decisive factor in mobile phone choice. While modes of acquisition depend on age, ownership of mobile devices far outweighs sharing. Learning of photo-visual and reproduction skills takes place horizontally through interaction with family and friends at school/work or within the family. There was no evidence of more cognitively demanding skills such as branching and information skills despite formal teaching. However, respondents are aware of risks online and appear to be security conscious. Reasons for becoming an online participant ranged from saving on communication costs to keeping in touch with distant relatives and friends.

Among the respondents, age proved to be a significant factor affecting the mobile Internet and digital acquisition experience. For older women digital literacy was mediated by formal or informal learning about computers, while younger ones were mobile-first users. Dwesa being a relatively poor area, issues of cost also played a significant role, e.g. in determining the type of applications used and activities performed. The rural condition was referred to indirectly in terms of distance from people, services and events.

The work described here illustrates the nuances of mobile devices and related skills acquisition. It also relates mobile uses with specific constraints of the rural area under consideration as representative of many African realities. Future research should focus on the differences between formal and informal learning in relation to mobile literacies. In particular, while basic use skills can be learnt from peer and by trial and error, more advanced and critical forms of literacy appear to require explicit teaching. Intergenerational transfer of skills also emerged as a potentially interesting topic for further exploration. The relatively small sample allowed for in-depth understanding but does not allow broad generalisation within the community, let alone across different contexts. Similar studies could be replicated among women in urban or peri-urban areas for comparison. While the research design we employed was suitable for an initial exploratory study, a direct comparison between genders and an exploration of gender dynamics in relation to mobile acquisition and use would provide a more comprehensive picture.

Appendix A: Interview Schedule

1. What type of mobile do you have?
2. How did you get it?
3. When did you get your current phone?
4. Why do you have that particular device?
5. What is the Internet for you?
6. Where (on which device) do you access the Internet?
7. When did you first access mobile internet/when did you get your first phone that had internet?
8. Why did you first access mobile internet?
9. How did you learn about mobile internet?
10. How did you learn to use the Internet on your phone?
11. How did you learn about the symbols and/or icons on your mobile?
12. How/Where do you search for information on the internet?
13. What do you use the Internet for? When last?
14. Do you share content online?
15. What content do you share? How do you share the content?
16. Do you create videos or memes?
17. What do you use the internet the most for? When last?
18. What do you use the internet the least for? When last?
19. As a woman what are some of the problems that the internet helps you solve on a daily basis?

20. As a woman what does it mean to have internet access on a mobile? How would you say it is different for men?

References

Adepetun A (2015) Africa's mobile phone penetration now 67%. The Guardian Newspaper Online. http://guardian.ng/technology/africas-mobile-phone-penetration-now-67/. Accessed 14 Mar 2016

Ahonen T (2008) Mobile as 7th of the mass media: cellphone, camera phone, iPhone, smartphone. Future text, London

Aker JC, Mbiti IM (2010) Mobile phones and economic development in Africa. Center for global development working paper. http://dspace.africaportal.org/jspui/bitstream/. Accessed 10 Mar 2016

Bosch T (2013) Youth, Facebook and politics in South Africa. J African Media Stud 5(2):119–130

Burrell J (2010) Evaluating shared access: social equality and the circulation of mobile phones in rural Uganda. J Comput.-Mediated Commun 15(2):230–250

Buskens I, Webb A (2009) African Women and ICTs. Investigating technology, gender and empowerment. http://ifap-is-observatory.ittk.hu/node/77

Buthelezi M (2015) Money-related uses of mobile phones in a South African rural area. Honours thesis. School of Journalism and Media Studies, Rhodes University

Castells M, Fernandez-Ardevol M, Qiu JL, Sey A (2007) Mobile communication and society: a global Perspective. The MIT Press, Cambridge

Cristoferi M (2015) Mediating distance: investigating and deconstructing mobile phone use in two marginalised South African contexts. Paper presented at SACOMM 2015, Cape Town, 28–30 September 2015

Collopen N (2015) An exploration of media and mobile usage ecosystems in marginalised areas: the case of Dwesa. Honours paper. School of Journalism and Media Studies, Rhodes University

Dalvit L, Alfonsi R., Isabirye N, Murray S, Terzoli A, Thinyane M (2006) A case study on the teaching of computer training in a rural area in South Africa. In: Proceedings of the 22nd Comparative Education Society of Europe (CESE) conference, Granada, 3–6 July 2006

Dalvit L, Kromberg S, Miya M (2014) The data divide in a South African rural community: a survey of mobile phone use in Keiskammahoek. In: Proceedings of the e-Skills for knowledge production and innovation conference 2014, Cape Town, 17–21 November 2014

Dalvit L (2015a) Beyond m-learning: mobile video narratives for teaching, research and community engagement at a South African university. Paper presented via teleconference at the Mobile Innovation Networks Australasia (MINA) conference, Melbourne, 19 November 2015

Dalvit L (2015b) Mobile phones in rural South Africa: stories of empowerment from the siyakhula Living Lab. In: Dyson LE, Grant S, Hendriks M (eds) Indigenous People and Mobile Technologies. Routledge

Dalvit L, Strelitz L (2013) Media and mobile phones in a South African rural area: a baseline study. In: Proceedings of the emerging issues in communication policy and research conference, University of Canberra, Canberra, 18–19 November 2013

Deen-Swarray M (2016) Toward digital inclusion: understanding the literacy effect on adoption and use of mobile phones and the Internet in Africa. Inf Technol Int Dev 12(2):29–45

Donner J (2008) Shrinking fourth world? Mobiles, development, and inclusion. In: Katz J (ed) Handbook of mobile communication studies. MIT Press, MA, pp 29–42

Du Plooy L, Zilindile M (2014) Problematising the concept epistemological access with regard to foundation phase education towards quality schooling. South African J Child. Educ. 4(1): 187–201

Eshet-Akalai Y (2004) Digital literacy: a conceptual framework for survival skills in the digital era. J Educ Multimedia Hypermedia 13(1):93–106

Esselaar S, Weeks K (2007) The case for the regulation of call termination in South Africa: an economic evaluation. ICASA and LINK Centre, University of the Witwatersrand, Johannesburg, South Africa

Fripp C (2014) South Africa's mobile penetration is 133%. HTXT Africa. http://www.htxt.co.za/2014/10/23/south-africas-mobile-penetration-is-133/. Accessed 26 Apr 2016

Goggin G (2006) Cellphone culture. Mobile technology in everyday life. Routledge, Abingdon

Ezemenaka E (2013) The usage and impact of Internet enabled phones on academic concentration among students of tertiary institutions: a study at the University of Ibadan, Nigeria. Int J Educ Dev Inf Commun Technol (IJEDICT) 9(3):162–173

Frost, Sullivan (2015) Sub Saharan mobile end-user trends. AFR_PR_SJames_MA20-65_06Jan15.pdf

Goldstuck A (2010) Internet access in South Africa 2010: a comprehensive study of the internet access market in South Africa. World Wide Worx, Johannesburg

Goldstuck A (2013) The many needs for Internet speed. http://www.gadget.co.za/pebble.asp?relid=6826. Accessed 26 Aug 16

Goliama CM (2011) Where are you Africa? Church and Society in the phone age. Langaa Research and Publishing Common Initiative Group, North West Region Cameroon

GSMA (2015) The mobile economy 2015. http://www.gsmamobileeconomy.com/GSMA_Global_Mobile_Economy_Report_2015.pdf. Accessed 17 Mar 2016

Gunzo F, Dalvit L (2012) A survey of cell phone and computer access and use in marginalised schools in South Africa. Paper presented at the 3rd international conference on mobile communication for development, New Delhi, 28–29 February 2012

Gunzo F, Dalvit L (2014) One year on: a longitudinal case study of computer and mobile phone use among rural South African youth. In: Steyn J, Van Greunen D (eds) ICTs for inclusive communities in developing societies. Proceedings of the 8th international development informatics association conference, Port Elizabeth, 3–4 November 2014

Hancock D, Algozzine B (2006) Doing case study research: a practical guide for beginning researchers. Teachers College Press, New York

Hilbert M (2011) Digital gender divide or technologically empowered women in developing countries? A typical case of lies, damned lies, and statistics. Women's Studies International Forum. http://www.martinhilbert.net/DigitalGenderDivide.pdf. Accessed 14 Aug 2016

Holden M (2012) Life with or without the Internet: the domesticated experiences of digital inclusion and exclusion. MSc in Media and Communications. Department of Media and Communications, London School of Economics and Political Science

ITU (2013) Study on international internet connectivity in Sub-Saharan Africa. http://www.itu.int/en/ITU-D/Regulatory-Market/Documents/IIC_Africa_Final-en.pdf. Accessed 11 Nov 2016

Kalba K (2008) The adoption of mobile phones in emerging markets: global diffusion and the rural challenge. Int J Commun 2:631–661

Kreutzer T (2009) Generation mobile: online and digital media usage on mobile phones among low-income urban youth in South Africa. University of Cape Town, Cape Town

Mapi TP, Dalvit L, Terzoli A (2008) Adoption of ICTs in a marginalised area of South Africa. Africa Media Rev 16(2):71–86

Moore A (2007) Mobile as the 7th mass media: an evolving story. An SMLXL White Paper

My Broadband (2015) South Africa's big smartphone Internet uptake. http://mybroadband.co.za/news/smartphones/127556-south-africas-big-smartphone-internet-uptake.html. Accessed 23 Dec 2016

Obijiofor L (2015) New technologies in developing societies: from theory to practice. Palgrave MacMillan, United Kingdom

Pade-Khene C, Palmer R, Kavhai M (2010) A baseline study of a Dwesa rural community for the Siyakhula information and communication technology for development project: understanding the reality on the ground. Inf Dev 26(4):265–288

Powell AC (2012) Bigger cities, smaller screens: urbanization, mobile phones, and digital media trends in Africa. http://www.meducationalliance.org/sites/default/files/bigger_cities_smaller_screens_urbanization_mobiles_phones_and_digital_media_trends_in_africa.pdf. Accessed 20 Jan 2017

Research ICT Africa (2012) Internet Going Mobile: Internet access and usage in 11 African countries. RIA policy brief no. 2. www.researchICTafrica.net

Rogers EM (2003) Diffusion of innovations, 5th edn. Free Press, New York

Roux K, Dalvit L (2014) Mobile women: investigating the digital gender divide in cellphone use in a South African rural area. In: Proceedings of the e-Skills for knowledge production and innovation conference 2014, Cape Town, 17–21 November 2014

Sey A (2011) New media practices in Ghana. Int J Commun 5(11):380–405

Statistics South Africa (2011) Eastern Cape Municipal Report. http://www.statssa.gov.za/census/census_2011/census_products/EC_Municipal_Report.pdf. Accessed 8 Feb 2016

Statistics South Africa (2015) Census. http://www.statssa.gov.za/. Accessed 8 Feb 2016

Thinyane M, Dalvit L, Terzoli A (2008) The internet for rural communities: unrestricted and contextualised. Paper presented at ICT Africa, Addis Ababa, 13–15 February 2008

United Nations (2005) Gender equality and empowerment of women through ICT. Women 2000 and Beyond. http://www.un.org/womenwatch/daw/public/w2000-09.05-ict-e.pdf. Accessed 8 Mar 2016

Yin RK (1999) Case study research. Design and methods. Sage Publications, London

The Potential Role of Digital Technologies in the Context of Forced Displacement

Andreia Ribeiro[✉] and Vania Baldi[✉]

Departamento de Comunicação e Arte, DigiMedia (CIC.Digital), Universidade de Aveiro,
Campus Universitário de Santiago, 3810-193 Aveiro, Portugal
{andreiar,vbaldi}@ua.pt

Abstract. In light of the European migrant crisis and considering the growing use of new technologies by humanitarian agencies in their operations, it is important to reflect on the possible social and technological convergence of these phenomena in an attempt to minimize the damages and respond to the needs of refugees. This research focuses on the adoption of new technologies in humanitarian aid, with emphasis on the use of smartphones by refugees who escape the various wars occurring in countries such as Syria or sub-Saharan areas and that are trying to successfully adjust to their host country.

Keywords: Refugees · Infocommunicacional strategies · Mobile apps · Humanitarian aid

1 Introduction

The widespread use of smartphones that has been happening in the last decade and the always increasing development and use of instant messaging and social networking platforms, are essential factors of a digital culture that manifests itself in a set of communication media in the daily life of individuals. This rapid spread of ICT is also changing the approaches and strategies used by organizations and individuals involved in situations of conflict and natural disasters. New technologies enable at-risk communities to quickly and easily send out requests and alerts and share important information with humanitarian aid agents.

Currently, populations affected by disasters or conflict find it easier to access information and many of the information needs that arise during a crisis can increasingly be answered faster thanks in large part to mobile technologies. The development of an humanitarian approach that is more geared towards new technologies is therefore essential and inevitable. New technologies, such as smartphones, SMS, georeferencing, social media and crisis mapping, enable affected communities to access, produce and share useful and actionable information. Consequently, many humanitarian agencies are starting to adopt these instruments in their initiatives, always taking into account the feedback provided by the communities.

This investigation, which is still under development, focuses on the features of some existing technologies used in humanitarian aid, and subsequently, we'll evaluate the combined uses of the various mobile apps and websites aimed at refugees through

Ó. Mealha et al. (eds.), *Citizen, Territory and Technologies: Smart Learning Contexts and Practices*, Smart Innovation, Systems and Technologies 80, DOI 10.1007/978-3-319-61322-2_11

interviews and focus groups with their users. We expect to come to an understanding of the features available in these apps and find out if these are appropriately tailored to the needs of its target audience. To this end, there will have to be an approach not only with refugees, but also with representatives of aid organizations, in order to investigate possible gaps that will allow us to develop a mobile application that takes these opinions into account.

2 Infocommunicacional Strategies to Enhance the Response to Forced Displacement

For this paper, we consider a refugee a person who is forced to leave his or her country of origin due to an armed conflict or persecution. Until they are registered with The UN Refugee Agency (UNHCR), displaced persons are not officially considered as refugees, as such they have no right to protection or assistance from organizations. The lack of safe access to humanitarians and numerous administrative and political factors stand out as the main obstacles to humanitarian support in Syria. The country is considered one of the most dangerous countries in the world for humanitarian or health professionals, and humanitarian facilities are deliberately targets from all groups involved in the conflict (OCHA 2016). It is opportune to use techniques and tools that refugees have access to and give them some agency so that they aren't merely dependent on humanitarian aid.

There are now more portable devices, especifically smartphones, than people (ITU 2016a). This ubiquitous access to these devices and to the internet is generating new ways for affected communities and humanitarians to organize and respond to obstacles and necessities. According to data from ITU (International Telecommunication Union), 80% of the global population uses smartphones e 33% accesses the internet through a handheld device. In a lot of developing and underdeveloped countries, internet access is made almost exclusively through mobile devices. As such, is very likely that communities affected by conflict or natural disasters produce and consume digital content (ITU 2016a).

The fact that refugees have smartphones in their possession has been one of the arguments used to discredit their situation, mistakenly assuming that they do not need help because they have money to purchase these devices. However, as prices of these continue to drop, along with a decline in mobile network subscriptions prices (ITU 2016b), these factors made them accessible to the most underprivileged individuals. The smartphone and the internet have become a vital part of a refugee's daily life. With these, they can access maps, public transport schedules or check social media for tips and real-time information on possible obstacles from other refugees (Plataforma de apoio aos refugiados 2016). Refugees in Syria use YouTube to share images of what is happening and use Skype and WhatsApp to get in touch with family and friends in their home country, to get information and to send requests to the humanitarian community (IFRC 2013). These tools have been instrumental for individuals in Syria to conduct themselves and deal with their problems more efficiently.

The omnipresence of smartphones has also contributed to a propensity of monetary remittances made on these devices, and currently, most of the donations to social causes are made through mobile phones and social media. There has also been an increase of these money transfers to communities in the diaspora, who transfer directly to friends and family in the countries of origin (Loh 2016). Thanks to smartphones, refugees are able to perform this task quickly and in a more efficient way. It is also an approach that has been successful with humanitarian agencies, with some giving out vouchers to be exchange for food and shelter (IFRC 2013).

A study conducted by Internews found that the information that refugees most value are news about their family in the country of origin, how to find work in the host country and information on food and housing (di Giovanni 2013). As for Wall et al. (2015), they learned that the most important information for refugees' day-to-day lives are news about personal contacts in the camps, in Syria and about the Syrian conflict and information on the aid programs available in the camps. The vast majority (63%) prefer to be contacted through mobile phone/SMS, while the rest are divided between preferring face-to-face contact with officials, or looking through the internet, television or newspapers as a means of obtaining information. In both studies, most refugees believe that it would be extremely helpful to receive information about aid services on their mobile phones and that the use of these devices is essential to their daily life. Only 14% of the participants said they did not have access to a mobile phone and 40% of those who do have one, have a smartphone. These are mostly used to make and receive calls and text messages, with instant messaging apps like Whatsapp being the most popular means of communication among all classes of refugees (di Giovanni 2013).

Communities can then more easily communicate and share important information with each other. Whether through this sharing of information or of clothing and food, diasporas and local communities, allied with technologies, are increasingly using a do it yourself approach, somewhat diminishing the humanitarian agencies' role (IFRC 2013). And aid agencies are no longer merely dependent on information shared by other humanitarian organizations and the media, but can also access information generated by the communities themselves. They too are generators of structured and georeferenced digital data in various formats, such as text, image, video and voice thanks to the use of geospatial technologies (IFRC 2013). Georeferencing technology recognizes the geographical position of the individual and this way, he/she will have access to more information about that place or specific elements of it via videos, sounds, texts, infographics, etc. These technologies allow its users to contextually learn the information and also contribute to add new material, suggestions and emotions to this territory (Oliveira and Baldi 2015). And so, this geographic information is no longer merely produced by official institutions and private mapping companies, but are also created by individuals who want to share their data with other users. Due to smartphones and mobile apps that use type of technology, the map as we know becomes a digital and dynamic one, an interface in which its content is shared and social networks are established (Baldi and Oliveira 2013).

Information and communications technology (ICT) create a continuous necessity of upgrade and adaptability of the tools that humanitarian agencies employ in their actions. ICT allowed them to create new registration and support systems that help them distribute their aid. Institutions, such as UNHCR, started using databases to determine the size and structure of a population. Instruments like this help aid agencies to coordinate their efforts and resources more efficiently. Organizations also use these new technologies to send out alerts, train volunteers, connect and engage with communities at risk and to raise awareness and monetary funds (Loh 2016).

A number of humanitarian organizations are experimenting with these new sources of information, digital technology, data collection and real-time monitoring platforms, such as the American Red Cross' Social Media Digital Operations Center, the first center for humanitarian aid actions based on content from social media. The center uses software (Radian6) to control and analyze social media in real time and allows the Red Cross to do a personalized and specific search to better respond to the needs of its users. And since the software allows up to 25 users, a number too high for the institution to permanently keep up, it followed the example of the DHN and trained volunteers on how to use the software (IFRC 2013). Radian6 is used as a tool to monitor and analyze mentions of a company, brand or keywords made on social media and also allows real-time interaction in the website where the remarks are taking place (Web Analytics World 2012). As such, it is mainly used by companies as a marketing tool, not having been developed with the aim of aiding humanitarians. And given that the humanitarian sector does not represent a financially attractive market for software development companies, they rarely maintain or upgrade humanitarian aid programs. Hence organizations have to resort to open-source software, or in the case of Radian6, costly options (IFRC 2013).

2.1 Crowdmap as Collaborative Intermediation

The vast majority of these aid tools rely on crowdsourced crisis mapping. This instrument consists of a map that encompasses the collection, visualization and analysis of data in real time during a crisis thanks to information acquired by mobile applications or websites, aerial and satellite images, etc. (Meier 2011). Standby Volunteer Task Force (SBTF) is one of the multiple initiatives of professional humanitarian networks that dedicate themselves to humanitarian aid online. The non-profit organization created a crisis map of Libya with content from social media and from the Office for the Coordination of Humanitarian Affairs (OCHA) which provided them with a list of indicators such as people displacement, health, logistics and threats. The map, password protected, allowed volunteers to check, analyze and verify vast quantities of information related to these categories. Later, a public map was also made available, however the information in this was anonymous and only disclosed with a 24 h delay for security reasons (SBTF & OCHA 2011). During a crisis, people need to have fast and easy access to information that its relevant to them and, such information, must be arranged intuitively and offer feasible data. And that's where these live maps come in and quickly facilitate the visualization of what is happening and where, improving the knowledge of the situation in question.

For Jesse Hardman and Jacobo Quintanilla, messages of support and information sharing as a form of assistance are often best delivered by individuals who share a link with both local communities and humanitarian agencies. The authors mention the case of Ramanan Santhirasegaramoorthy, a radio host with a show in Sri Lanka aimed at people displaced by war. The host provided news and information to individuals about where to find basic resources, how to stay safe, and how to get in touch with humanitarian and government agencies. After learning about the show, Internews trained Santhirasegaramoorthy and his team, teaching them humanitarian principles, how to do disaster coverage, how to work with the government, military and humanitarian organizations and how to connect and interact with listeners in need of help. Currently living in Canada, Santhirasegaramoorthy now hosts a radio show aimed at the Tamil diaspora living in the country. In his new platform, he focuses on issues related to integration, sharing tips on how to adjust to life in Canada and how to deal with the stress of living and working in a Western society (IFRC 2013).

2.2   Networks of Reliability to Filter News Online

This easier access to content creation and sharing allows individuals to self-organize and help others at risk. However, this also prompts for an overwhelming amount of information online that makes it harder for both organizations and individuals to filter it. Hence the importance of initiatives such as the Digital Humanitarian Network (DHN). DHN was designed "to provide an interface between formal, professional humanitarian organizations and informal yet skilled-and-agile volunteer and technical networks" (Digital Humanitarians 2015). The members of this network are diverse, from large corporations to small and medium-sized non-governmental organizations, and each has a multiplicity of skills among them, cartography, social media monitoring, technology development and data analysis. Launched in 2012, DHN has already provided information, imaging, mapping and technical crisis development to organizations such as OCHA and Doctors Without Borders (Digital Humanitarians 2015). One of the interfaces that has been developed is the "Services Advisor" that links refugees to the humanitarian agencies they need most. The interface aims to facilitate the sharing of information about humanitarian services available in Jordan and to enable aid organizations and governments to communicate with each other. Its contents are updated weekly by humanitarians on the ground and can be filtered by type, location and proximity to the service (PeaceGeeks n.d.).

As with traditional media, social media also act as a critical medium in the spreading of information during conflict. However, the vast amount of information available in these makes it difficult to filter the most important and relevant content. As so, it is appropriate to use tools that simplify and verify this research, as a way to accelerate this process. Thus the initiative of the Federal Emergency Management Agency (FEMA) during Hurricane Katrina, that created a "rumor control" site listing all the rumors circulating about the disaster and labeling which ones were true and false. As for Twitter, the social network is the online tool most used to share information during disasters and conflict. This audience preference comes mainly from the hashtags that this social networking site has popularized and that make it easy to categorize, find and participate in conversations on a particular topic (Hashtag Definitions n.d.). During typhoon Pablo, the Philippine Government regularly used its Twitter account to make status updates and created specific hashtags so that people could keep track of the situation (#PabloPH) and also to ask for help (#reliefPH and #rescuePH) (IFRC 2013).

The use of new technologies in humanitarian action offers concrete ways to make aid more efficient, accountable and transparent. However, it is worth noting that in a lot of affected areas, in addition to scarce information, there might be limited mobile network coverage, restricting both the population and humanitarians of access to the internet. Therefore, to reach the highest numbers of individuals possible, there has to be a balance between the use of tradicional media, such as newspapers and radio, and new technologies. Instead of trying to figure out which of these is the most effective, since literacy and digital literacy is going to differ from area to area. Aid humanitarian agencies should keep in mind which information is the most relevant, which channels those individuals use and trust and how they communicate with each other and with other communities (IFRC 2013).

3 Final Considerations

For the purpose of this short paper, we've summarised our listing of the state of the art of the mobile applications for refugees to the three we found more interesting and relevant to the product we're trying to develop. "RefInfo" is a mobile app, available for Android and iOS, intended for refugees arriving in the Netherlands. With this app, the creators hope that refugees are able to find all the information they need in one single place. Its menu consists of: a 'news' section with links to Facebook pages of media and aid institutions; a 'next step', with numerous categories with information regarding legislation, accommodation, how to open a bank account, insurance, SIM cards, integration, among others; an 'about the Netherlands' section about the country, its customs, culture and history; a division dedicated to the learning of Dutch, with multiple YouTube videos and sound clips to learn the language; a section with dutch cartoons aimed at children, mainly for keeping them entertained in the waiting queues; and in 'locations' we find several Embassies and Migration and Foreign Services contacts. All its menus and contents are in English and Arabic.

"Helphelp2" is an app that connects organizations to volunteers who want to help. When opened, the application shows a map of the area where the user is with the places he/she can go to provide help, like making a donation, for example. Each location on the map is identified with the name of the institution, opening hours, distance, address and items they would like to be receive from donors. The vast majority of these data are centered in Germany and the remaining few in Austria.

Also taking advantage of the georeferencing technology, it's the Android app "Refugee Aid". Aimed at refugees and citizens who want to help them, the application works as a service and goods exchange platform, proving information on offers around the area where the user is. When logged in, volunteers can create offers with whatever they would like to provide, like clothes or housing, or a simple invitation for a meal, always stating a contact and location. And the refugees, can put up ads asking for something that they need.

As previously mentioned, this paper is the first step in an ongoing investigation, where the final goal will be to develop a mobile application that can help refugees with their integration in Portugal. In this initial phase, we strived to make an approximation to the subject of refugees and new technologies by listing pertinent literature and some mobile applications that support refugees. Later on, qualitative data will be collected from refugees and representatives of humanitarian organizations in order to generate an in-depth understanding of the uses, trends and potential of mobile applications in migratory diasporas. In our final product, we hope to comprise not only the invaluable informative nature of apps like "RefInfo", but also the social component present in "Refugee Aid" and "helphelp2" that we believe is vital for better integrating refugees in a context with such different languages and lifestyles than what they were used to (Fig. 1).

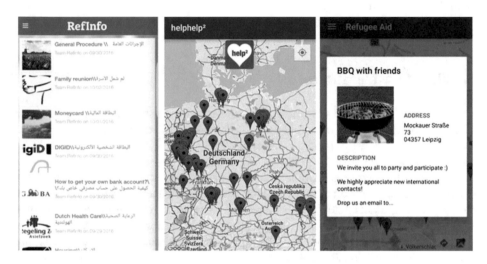

Fig. 1. Screenshots for the apps RefInfo, helphelp2 and Refugee Aid, respectively.

References

Baldi V, Oliveira L (2013) Território hipermediatizado e convergencias multilocalizadas: dialética entre terra e núvens. Experiências de Consumo Contemporâneo, December 2013, pp 28–46. https://doi.org/10.13140/RG.2.1.5077.7128

di Giovanni J (2013) Lost: syrian refugees and the information gap. Internews. http://www.internews.org/research-publications/lost-syrian-refugees-and-information-gap. Accessed 4 Nov 2016

Digital Humanitarian Network (2015) DHN catalogue. http://digitalhumanitarians.com/resource/dhn-catalogue. Accessed 20 Jan 2017

Hashtags. https://www.hashtags.org. Accessed 15 Jan 2017

IFRC (2013) World disasters report: focus on technology and the future of humanitarian action. Int Federation Red Cross Red Crescent Soc. doi:10.1111/j.0361-3666.2005.00327.x

ITU (2016a) ICT facts and figures 2016. Geneva

ITU (2016b) Measuring the information society report. Geneva. ISBN: 978-92-61-21431-9

Loh T (2016) Digitizing refugees: the effect of technology on forced displacement. Gnovis J. http://www.gnovisjournal.org/2016/04/29/digitizing-refugees-the-effect-of-technology-on-forced-displacement/. Accessed 4 Nov 2016

Meier P (2011) iRevolutions. What is crisis mapping? An update on the field and looking ahead. https://irevolutions.org/2011/01/20/what-is-crisis-mapping. Accessed 25 Oct 2016

OCHA (2016) 2017 Humanitarian needs overview. http://hno-syria.org. Accessed 3 Jan 2017

Oliveira L, Baldi V (2015) O potencial educativo do território hipermediatizado_: dos lugares do conhecimento ao conhecimento coproduzido nos lugares. prisma.com 28:65–85. http://revistas.ua.pt/index.php/prismacom/issue/view/257

PeaceGeeks. https://peacegeeks.org/products/services-advisor. Accessed 20 Jan 2017

Plataforma de apoio aos refugiados (2016). Factos e argumentos para desfazer medos e mitos: refugiados. http://www.refugiados.pt/teste. Accessed 20 Jan 2017

Plataforma de Apoio aos Refugiados. Mitos & Medos. http://www.refugiados.pt/a-crise-dos-refugiados/mitos-medos. Accessed 20 Jan 2017

SBTF & OCHA (2011) Libya crisis map deployment. http://www.standbytaskforce.org/2011/09/01/libya-crisis-map-report. Accessed 15 Jan 2017

Wall M, Otis Campbell M, Janbek D (2015) Syrian refugees and information precarity. New Media Soc. doi:10.1177/1461444815591967

Web Analytics World (2012) Radian 6 overview. http://www.webanalyticsworld.net/analytics-measurement-and-management-tools/radian-6-. Accessed 20 Jan 2017

Supporting Participatory Citizenship Insights from *LXAmanhã* Platform

Raissa Sales[1,2], Ana Carla Amaro[1,2(✉)], and Pedro Amado[1,2]

[1] University of Aveiro, Aveiro, Portugal
{raissakaren,aamaro,pamado}@ua.pt
[2] Digimedia (CIC.Digital), University of Aveiro, Aveiro, Portugal

Abstract. Participation in decisions surrounding city-related interests promotes transparency of governance, enables progress, democracy and enhances the empowerment of citizens as central elements in a political, economic and cultural context. The purpose of this paper is to highlight the relevance of online digital platforms to enhance participatory citizenship in the current social, economic and political fabric of cities, through the analysis of a case study. In this way, this paper focuses on the case study of the *LXAmanhã* platform, as part of an on-going research on online platforms to support participatory citizenship. The design and functionality of the platform were analysed, as well as the data related to the projects submitted between 2012 and 2016. The results translate into insights and thoughts regarding citizens' participation in the platform, platform mechanisms and features and technological resources for the future. However future work is still needed, it was possible to conclude that the main challenge for these platforms is sustaining the community participation and significant engagement.

Keywords: Participatory · Citizenship · Culture · Digital platforms · LXAmanhã

1 Introduction

Citizens' participation in decisions surrounding city-related interests enables progress and democracy and, in addition to fostering common discourse, also enhances the formation and development of people as central elements in a political, economic and cultural context. Nowadays, digital platforms and environments are increasingly used to support the manifestations of the participatory citizenship and for the debate and organization of social and societal transformations. In some cases, citizen participation happens passively, as a result of requests from government agencies, non-governmental organizations, associations, syndicates or private companies. Accessing online public services, viewing parliamentary proposals, seeking information on transport, services and cultural activities are simple and common tasks, available through various tools and platforms.

Participatory culture, especially in the current context of Web 2.0, asks citizens for active and voluntary social action, which should consider the collective commitment to fundamental decisions that affect life in society, and that goes beyond simply clicking to search the web or comment posts. In the context of cities, participatory citizenship

© Springer International Publishing AG 2018
Ó. Mealha et al. (eds.), *Citizen, Territory and Technologies: Smart Learning Contexts and Practices*,
Smart Innovation, Systems and Technologies 80, DOI 10.1007/978-3-319-61322-2_12

can be seen as a form of participation in possible improvements to the structures of urban environments and the functioning of public equipment and services, as well as in the subsistence of cultural and educational projects, among other possible examples.

Without neutralizing or demeaning other types of resources or ways of discussing or practicing citizenship, mediation of technologies emerges as a way to foster these discussions and practices, allowing the use of digital online support to disseminate and expand actions and projects and thus achieve better results, such as increased citizen compliance, greater diversity of resources, more attractive tools and lower cost.

The *LXAmanhã* platform is one of these online spaces for the exercise of participative citizenship, encouraging citizens to research, share and collaborate online to improve the urban spaces of the city of Lisbon, as well as the enjoyment of these spaces by citizens. This paper presents the study of the particular case of the *LXAmanhã* platform, a work that is part of a survey and analysis of a set of platforms to support participatory citizenship. First, this research contributes is to the existing body of knowledge by providing an objective analysis of a local case study features and participation data. Second, it is expected that the study will allow to identify a set of good practices that can be enunciated as guidelines for platforms to be developed in the same scope, to encourage the active and meaningful participation of the citizens.

2   Background

In a network, individuals combine and share their knowledge, references and culture. In this process, what is shared can be corrected, opened, processed, enriched and evaluated, something that is fundamental to what Levy (2004) called "collective intelligence". For the author, communities function as intelligent filters that help deal with the excess of information and unify the alternative views of a culture. By disregarding the specificities of situations and individual needs, networks are more efficient (Levy 2004). In this way, the existence of a collective is essential for mobilization, organization or any interaction in a cultural construction and development perspective. Jenkins (2009) emphasizes that this construction and development do not occur through devices, however sophisticated they may be, since they emanate from the subjects and their social interactions. It's important to understand about a participatory culture.

Jenkins et al. (2016) consider that any element developed for the exercise and promotion of a participatory culture, such as, for example, a digital platform, cannot limit access to the cultural means of production and circulation, nor fragment and isolate the public rather than provide opportunities to create and share culture and build hierarchies, or even hinder significant influence over the fundamental decisions that affect life in society. Participatory culture requires subjects' actions to be motivated and nurtured by an ethos of 'doing it together', in addition to 'doing it alone' (Jenkins et al. 2016). Moreover, another way of conceptualizing the term and conceiving its development is to understand it as something dynamic and unfinished, defined in the current context and in parallel with institutionalized structures and powers. This culture is produced by people who find voice, agency, and collective intelligence in spaces on Web 2.0 platforms (Jenkins et al. 2016).

Shirky (2011) points out that the simple fact of creating something online together with other people and then sharing it represents the repercussion of the old model of culture but technology enhanced. For the author, the concept of participatory culture refers to the experiences lived by the subjects, transposed to and shared mainly on the web, since users take the offline knowledge to the online lives. The context, environment, sensations and social relationships online and offline can be integrated, and technologies can enhance this integration, adding possibilities to users and offering tools for group organization and expression.

In the culture of participation, citizens develop skills, if necessary, and systematize communicational processes and actions in favour of resolutions for causes, socio-environmental support, political manifestations, economic gains, entrepreneurial initiatives, among others. They empower themselves through interaction, initially triggered by an exchange of information to produce knowledge, needs and desires, and they do so not only through the appropriation of media artefacts, but also in digital platforms organized for several purposes.

Jenkins et al. (2016) compare the effort of citizens in the context of participatory citizenship to the performance of fans, an audience that plays a key role in the production of media content. Like the fans, active citizens take a deep commitment, but with civic responsibility, often disinterested for profit and driven only by the will and social relevance of collaboration. Understanding the contributions of participatory culture involves an evaluation of practices, which, according to Jenkins et al. (2016) has also allowed the realization that the greatest social transformations happen through a shared vision of how a better society could be.

Individuals post on blogs, social networking groups, complaint and opinion pages, video sharing sites and other online services and platforms, multimedia content about their experiences in the city, in their convivial environments, in work contexts, between others. In the community, other members who have experienced similar or adverse experiences collaborate by publishing other content, new or complementary. Besides these, other possibilities open up to collaborate and produce in a network, so that the rights and duties of the citizens are transformed through a participatory culture lived in the cities.

In Portugal, some platforms approach this purpose with different strategies. The platform *Ideias à moda do Porto*, for example, is an initiative of a group of Porto city's citizens, which receives proposals to improve urban spaces, puts suggestions to vote and tries to promote some of them among the population. The *LXAmanhã* was create with the exactly same goal, but receives proposals for the city of Lisbon, inserted by the citizens directly into the platform, and then organized and showed using geolocation. Moreover, it is possible for users to support a proposal. Similarly, but supported by private entities, the platform *Por Um Bairro Melhor* allows the registration of citizen-owned projects to be carried out in the space of the neighbourhoods, and has a network of business partners to finance the selected projects. These examples have in common the fact that they operate based on the active participation of the citizen to promote the reflection and/or the execution of alternative forms of experiencing the urban spaces. They also promote the exercise of citizenship anchored in citizens' social commitments and, as a result, encourage a culture of participatory citizenship.

In the scientific context, the concept of participatory culture has become an interest of many areas and has been studied, mainly, in the political, administrative, cultural, communicational and media contexts. Over the past five years, some approaches have gained more prominence. The production of content with citizen participation and the media as a tool for social empowerment in manifestations, activism and democratic exercise (Linders 2012; Keller 2012) are much discussed topics. Another question of interest has been to understand how media and digital cultures allow citizens to get organized, to get involved and to act on collective issues and engage in the co-creation of the social fabric and the construction of the city shape (Lange and Waal 2013).

Research has also focused on the use of platforms in education, literacy and the exercise of citizenship in community, in order to encourage competences that allow more active participation in society (Kassam 2013; Taylor et al. 2012). Studies have also been carried out to generate reflections on government practices and reforms, especially regarding models for governance and administration practices (Nam 2012; Bonsón et al. 2012; Ellison and Hardey 2014).

3   Methods

The research this paper presents has a qualitative nature and is based on the case study (Gray 2014) of the *LXAmanhã* platform. This platform emerged in 2012 from the partnership of three colleagues from different areas and was inspired by the German project NextHamburg. It is an independent platform for collecting and consulting citizens' ideas to improve the city of Lisbon and reflect on its future, which allows the insertion of urban intervention proposals, which are mapped according to the locations to which they refer. The insertion of proposals, as well as access to other functionalities - such as, receiving notifications when comments are made on the proposals, post comments and collaborating on the development of suggestions, supporting ideas, etc. - depend on a registration as a platform user, by inserting personal data, a user name and a password. The support for an initiative is made through a voting system, by clicking a like-type button. After the submission, the idea becomes visible on the map, opening the possibilities for collaboration. The user may browse through the proposals by region of the city and by categories (represented with different colours), such as environment, architecture, social affairs, commerce, culture, public space, mobility, and others. The platform, by retaining and classifying solutions sought and co-opted by citizens, gives the community and policy makers a comprehensive vision of future needs and needs, and may therefore help elect initiatives for implementation.

The analysis focused on virtual documents available in the platform or through it, like texts, the platform chronology, the *LXAmanhã* blog, podcasts from radio programs, and all the project's proposals made available from 2012 to 2016. Proposals were analysed considering the insertion date, the amount of support achieved and the categories in which they were indexed by the citizens, when the project was inserted in the platform. Some features of the platform were also analysed, given its importance for the interaction process, such as the functionalities related with the insertion and visualization of the project's proposals, feedback and information updating. The platform's interface was not considered in the analysis. Although the study was based solely on data collected on the platform, *LXAmanhã* is also present through a blog and profiles on Facebook, Instagram and YouTube.

4   Results and Discussion

The results translate into insights and thoughts organized in two main topics: the frequency of participation in the platform during the period under analysis and the structure and functionalities of the platform. Considering the results of participation in *LXAmanhã* as a vulnerability of this type of online participation model, and based on the reflections and insights that resulted from the study, some improvement actions are suggested for the *LXAmanhã* platform and for other platforms or initiatives that come to be developed within the framework of participatory citizenship. It is also relevant to clarify that the interests of those responsible for the platform, whether economic or otherwise, were not considered or researched.

4.1   Participation in the Platform

The available map on the *LXAmanhã* platform allows citizens to view the proposals submitted by users since the platform's launch date, according to the area of the city of those proposals. The survey took into account the period of the last five years and all the categories that the platform integrates, in a total of 129 posts from 64 unique users, and 1010 supports. This computes to an average of 2 proposals per user and a support of 7,8 per proposal, which according to the online participation models (Preece and Shneiderman 2009) is an interesting participation average. Active participant users can be organized into 3 categories: one time participants, returning participants and regular participants. The first is a user that posts one entry and usually does not interact. The second is a user that posts from 2 to 4 entries, on the same month, around one or two related topics, and that interacts sporadically through support or comments. The last is a regular user that posts throughout the different years and interacts regularly with other users on different topics, promoting the platform and the causes. There are two clear examples of the latter: Jozhe (submitting proposals in all 8 categories between 2013 and 2016) and Luis (in 5 between 2012 and 2013). Although April seems to be the most active month for proposals, and Public Space and Mobility the most active topics, there were no specific trends detected. However, a decrease in these participations was observed (Fig. 1), especially from 2014, when four of the categories did not receive any project. Thus, this decrease in participation seems to constitute a vulnerability of this platform, and it is necessary to understand the possible reasons for having stopped attracting participants.

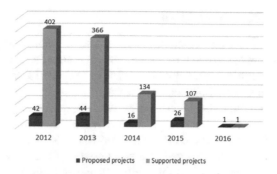

■ Proposed projects   ▥ Supported projects

Fig. 1. Number of supporters and platform posts.

Recalling the political and economic context during the period under review, and specifically in the year in which participation began to decline, the country was beginning to recover from a deep economic crisis. In these contexts, and although the sense of collectively and the interest of citizens in the common good may be even more necessary to overcome adversity, it is common for people to focus their efforts on individual and, at most, household survival.

Regarding the distribution of the proposals by the categories listed in the platform, it can be observed that, in the period under study, the categories with the most submissions are "public space", "mobility" and "environment", with 45, 34 and 23 posts, respectively (Fig. 2). Although the identification of the categories with the greatest interest for the users may be important to guide proposal's raising and disseminating efforts, an analysis of each of the projects submitted in the different categories allowed to conclude that the categorization, made by the users at the time of submission, does not always describe or correspond to the type of proposal. It was possible to find ideas of a more cultural or educational nature, categorized, for example, in "public spaces".

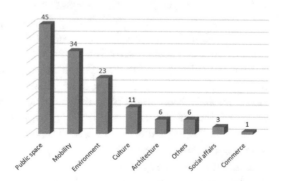

Fig. 2. Number of platform's posts per category.

It is also worth emphasizing that the data available on the platform, such as those previously reported, could be essential to feed an annual report available to citizens and stakeholders, including public agencies, research institutions, private companies, associations, NGOs, among others.

4.2 Platform Mechanisms and Features

It can be argued that citizen participation through the submission of projects contributing to the collective sphere has great significance, since it is an individual initiative that will impact a larger number of citizens. But, in the case of *LXAmanhã*, the platform becomes obsolete by not having tools that integrate the users.

In *LXAmanhã*, the contact between the users is made, exclusively, through comments that can be inserted in the proposals. In this sense, the focus on "doing it together" (Jenkins et al. 2016) is lost and therefore the opportunity to integrate such users to put the actions into practice is also wasted. It is not a matter of believing that the resources will determine the results, because this will only happen through the actions of the subjects. Cross-category opinions such as "necessity to transform the Garett Street into walkable paths and eliminate parking…" (mobility comment in a 2012 Public Space category proposal[1]), or "Location, for me, is very important, as the scenery is very beautiful…"[2], constitute such examples, expressed by users. Although the most active users interact and make efforts to mobilize and gain critical mass around topics and issues, what is proposed is to use the possible tools as strategies to extend the interaction between the users, to generate an easier, and richer involvement and commitment.

On the other hand, and by collecting information of interest to the public authorities, the platform could make the participation process even more expressive by extending it to other audiences, whether governmental, community or private, constituting itself as a participatory citizenship ecosystem, capable of mobilizing and being mobilized to transform ideas into actions.

A participatory ecosystem in the context of citizenship can be interesting to keep a financing relationship with public, private or alternative initiatives. Also, following the logic of crowdfunding platforms as a solution to broaden the power to fund proposals can stimulate contributions' submission and make them more effective. The projects that were to be implemented and the partners willing to carry them out or finance them would be recognized by a certification or seal *LXAmanhã*, a kind of recognition for collaboration and an evaluation in the field of participatory citizenship.

As for the page design in which the proposals are viewed, although it is simple and clean, it presents a static environment, attractive only by the images and map. Moreover, the lengthy texts and lack of feedback on the completion or progress of the proposal limit the users' knowledge to the point of generating disinterest.

It is also possible to verify that *LXAmanhã*'s blog and social networks could be better used in strategic communication actions with the public, deserving periodic updates, which currently do not happen. Through social networks, it would be possible to create interaction dynamics with users, such as the debate of ideas, voting, etc. It would also be possible to generate gamified challenges and to disseminate diverse multimedia information, such as videos or podcasts, to promote the platform, its goals and new projects.

---

[1] Available online at: http://www.lxamanha.pt/userpost/pedonalizacao-da-rua-garrett/.

[2] Available online at: http://www.lxamanha.pt/userpost/andar-de-bicicleta-livremente-junto-ao-rio/.

4.3   Technological Resources for the Future

It is also possible to verify that, presently, technological resources are available that, if properly employed in the case of the *LXAmanhã* platform, could boost the use of the platform and breathe new life into the project. Projects that can contribute to create and maintain a system in a predictable balance and help resolve conflicts (Stimmel 2015) are crucial in today's "smart" oriented environments. Taking advantage of the "mobile era" possibilities to renew and improve the user experience with *LXAmanhã* is one of the suggestions. Expansion to a Mobile App would enable on-site georeferenced publications. It would also be easier to collect and upload information such as photography, video and audio, for example, with testimonials from individuals who visit the places for which the proposals were launched. Augmented Reality could also be an interesting feature in this App. By adding a layer of virtual information to reality, the user could be notified regarding proposals submitted or in progress in a geo-localized way, being able to obtain more information by reading QR codes with the smartphone.

5   Conclusion

The results of the study confirm that the association between the interest in finding solutions, exploiting collaborative participation and using the mediation of information and communication technologies can be fundamental to the process of civic culture.

Although the results of this analysis are limited and cannot be extrapolated, the *LXAmanhã* case study can be articulated in the sense of projecting civility, establishing cultural influences of collaboration, sharing knowledge about the city and its improvements and facilitating communication between citizens and in society for decisions to be made.

The analysis of the participation data gathered throughout the time span of this project, led us to conclude that to contribute to a better performance of the platforms that follow these purposes and understandings, it is more efficient to privilege collective participation and invest in interaction between the subjects. The collaborative construction of improvement actions and the possibility of reaching other stakeholders of interest for the formation of a participatory ecosystem are ideal. As a result, we propose that, in order to legitimize the efforts sought by the platform and to achieve more concrete results regarding the implementation of the proposals, it is advisable to use partnership strategies with public, private and independent entities. By trying to implement these recommendations into an actionable plan or framework for specific projects, opportunities for execution and financing of proposals are enhanced, as well as citizen engagement.

The use of mobile technologies and mechanisms that facilitate the use of the platform can favour the dynamics, practicality and agility of using *LXAmanhã*, as well as promoting in users the feeling that the goals and contributions that feed these platforms are part of their day-to-day, to the point of being a reference for their performance as citizens.

Finally, it is proposed to use the data registered in these platforms to analyse or predict scenarios of need for citizens and the city and to direct efforts to strengthen the participatory culture in this area. In future studies, it will be appropriate to extend the analysis to other platforms by comparing them, as well as interviewing users in order to understand their interpretation of the use and relevance of participatory citizenship through these platforms.

References

Bonsón E, Torres L, Royo S, Flores F (2012) Local e-government 2.0: Social media and corporate transparency in municipalities. Gov Inf Q 29(2):123–132. doi:10.1016/j.giq.2011.10.001

Ellison N, Hardey M (2014) Social media and local government: citizenship, consumption and democracy. Local Gov Stud 40(1):21–40. doi:10.1080/03003930.2013.799066

Gray DE (2014) Doing research in the real world. SAGE Publications, London

Jenkins H (2009) Cultura da convergência. Aleph, São Paulo

Jenkins H, Ito M, Boyd D (2016) Participatory culture in a networked era: a conversation on youth, learning, commerce, and politics. Polity Press, Malden

Kassam A (2013) Changing society using new technologies: Youth participation in the social media revolution and its implications for the development of democracy in sub-Saharan Africa. Educ Inf Technol 18(2):253–263. doi:10.1007/s10639-012-9229-5

Keller JM (2012) Virtual feminisms: girls' blogging communities, feminist activism, and participatory politics. Inf Commun Soc 15(3):429–447. doi:10.1080/1369118X.2011.642890

Lange M, Waal M (2013) Owning the city: new media and citizen engagement in urban design. First Monday 18(11). http://dx.doi.org/10.5210/fm.v18i11.4

Levy P (2004) A inteligência coletiva: por uma antropologia do ciberespaço. Loyola, São Paulo

Linders D (2012) From e-government to we-government: Defining a typology for citizen coproduction in the age of social media. Gov Inf Q 29(4):446–454. doi:10.1016/j.giq.2012.06.003

Nam T (2012) Suggesting frameworks of citizen-sourcing via Government 2.0. Gov Inf Q 29(1):12–20. doi:10.1016/j.giq.2011.07.005

Preece J, Shneiderman B (2009) The reader-to-leader framework: Motivating technology-mediated social participation. AIS Trans Hum-Comput Interact 1(1):13–32 http://aisel.aisnet.org/thci/vol1/iss1/5

Shirky C (2011) A cultura da participação: criatividade e generosidade no mundo conectado. Zahar, São Paulo

Stimmel CL (2015) Building smart cities: analytics, ICT, and design thinking. CRC Press, Boca Raton

Taylor N, Marshall J, Blum-Ross A (2012) Viewpoint: empowering communities with situated voting devices. In: Proceedings of CHI 2012, pp 1361–1370. http://dx.doi.org/10.1145/2207676.2208594

Considerations on Information and Communication Overload Issue in Smart Cities

Joao Batista[1(✉)] and Rui Pedro Marques[2,3]

[1] CIC.Digital/DigiMedia, ISCA-University of Aveiro, Aveiro, Portugal
joao.batista@ua.pt
[2] ISCA-University of Aveiro, Aveiro, Portugal
ruimarques@ua.pt
[3] Algoritmi, University of Minho, Guimarães, Portugal

Abstract. This paper addresses the issue of information and communication overload in the context of smart cities. However this issue has been widely studied and its negative effects on individuals, organizations and societies are already known, as well as the multiple and diversified causes and solutions which have gradually been identified, the literature shows that it has been neglected in the context of smart cities. This paper approaches the subject of smart cities and mentions some contributions found on the literature to face the information and communication overload. The main concepts of information and communication overload are also described. Then, some measures are proposed to identify, prevent and deal with the information and communication overload in smart cities.

Keywords: Information · Communication · Overload · Smart city

1 Introduction

Smart cities are related with the idea of developing and implementing products and services that improve citizens' quality of life. Also, urban management of those products and services is expected to be more efficient.

Information and communication technologies (ICT) are key facilitators to the development and implementation of those products and services, and provide citizens, organizations and the urban society with some very important tools to achieve a better life in modern cities. However, now it seems clear that the adoption of those technologies may create situations of being overwhelmed by all the information and communication processes that are inherent in our daily lives, resulting frequently in situations of information and communication overload (ICO). The ICO issue has been researched for long, but it seems to be increasingly relevant because technological advances come with new challenges and problems.

The fact that smart cities are heavily depending on ICT suggests that ICO may be a serious issue, however the literature does not frequently approach this issue on the context of smart cities. In this paper, we suggest some suitable measures which may help coping with ICO in smart cities.

© Springer International Publishing AG 2018
Ó. Mealha et al. (eds.), *Citizen, Territory and Technologies: Smart Learning Contexts and Practices*,
Smart Innovation, Systems and Technologies 80, DOI 10.1007/978-3-319-61322-2_13

2 Smart City

The smart city concept has been frequently invoked over the past few years in situations related to the development of urban and regional contexts by exploiting ICT to create products and services which increase the quality of life of citizens and their communities.

However, the lack of a clear definition of the smart city concept and the need to use it in cross-cutting contexts leaded to the appearance of some clarification proposals. For example, Neirotti *et al.* (Neirotti et al. 2014) state that there is no general consensus on the smart city concept, although they verify that the demand and the systematic use of solutions involving ICT is a characteristic usually associated with this concept. These authors also acknowledge that *"those cities that are more equipped with ICT systems are not necessarily better cities"*, suggesting that other aspects than ICT need to be addressed in smart cities, such as urban planning.

Giovannella (Giovannella 2014) uses Maslow's pyramid (Maslow 1943) to describe what is a smart context highlighting the role of the individual and human capital as part of a context and taking into account their own needs. In the words of the author, *"a smart context is a context where the human capital (...) owns not only a high level of skills, but is also strongly motivated by continuous and adequate challenges, while its primary needs are reasonably satisfied"*. It is interesting that no technological elements are mentioned, highlighting the importance of the individual and relegating the issue of technology to an instrumental plan, but technology is implicit.

The domains of activity of smart cities are identified and systematized in a taxonomy proposal (Neirotti et al. 2014) with 6 main areas: *natural resources and energy, transport and mobility, buildings, living, government,* and *economy and people*. It becomes clear that the concept of smart city is extended to most urban human activities. This comprehensive taxonomy reinforces the idea that individuals should be placed at the center of the discussion of smart cities as well as the human capital they represent and the activities in which they participate.

However, the creation of products and services designed to promote high levels of efficiency and effectiveness in the processes of the aforementioned areas draws on many technologies, and in particular ICT. The use of technologies also promotes high levels of quality of life and active participation of individuals in urban communities. Nevertheless, this increasing and systematic use of technology, and especially ICT, is currently generating a huge volume of information and a density of communication processes that can be difficult to manage by individuals.

Some present relatively futuristic positions on the increasing use of ICT by individuals in their personal, organizational, or social activities, as the visionary anticipation presented by Matsuda (Matsuda 2016). This video shows a permanent, intrusive and alienating use of ICT, which seems to pose problems like states of confusion, stress, constant interruption, alienation from reality, and even identity problems. This permanent and intensive use of ICT seems to provoke our partial alienation from the real physical space that surrounds us, and does not seem to contradict the idea of a liquid society described by Zygmunt Bauman (Bauman 2013). In fact, instead of bringing people together based on real interactions among citizens to develop their life paths, there is a risk of developing contexts, such as smart cities, where the quality of life is increasingly based on ICT, possibly preventing individuals and communities from interacting and thus creating some kind of interdependence. Uncertainty will be increasingly present and, if an individual happens to be in a situation of fragility and cannot possibly be supported by a real and physical environment, then they will depend only on a network of contacts and communications, which may not ensure their quality of life and, in the limit, their survival (Bauman 2013).

It is also fair to acknowledge that many of the features shown in Matsuda's video (Matsuda 2016) are in general desired by citizens. They are information and communication technological solutions that help to solve some individual, organizational and societal issues. However, the trend seems to be combining many of these features into overlapping layers of information and communication, eventually in informational and communicational pollution.

Then some questions may arise. How can we collect, storage, and process data in a smart city? How can we identify the existence of redundancies? What will data be collected for, and how will that data collection be used to produce useful information in order to increase the quality of services? Are we getting to the point of finding that we have too much data and that all sorts of activities are recorded in a multitude of possibly redundant and difficult-to-perceive registers? Will all this data and information flow into networks of communication and of services that can lead to situations of hyper-reality such as that presented by Matsuda (Matsuda 2016)? If the trend continues to evolve into increasingly saturated information and communication environments, coupled with increasing levels of uncertainty about citizens' livelihoods, how can we ensure that smart cities will be inclusive?

The above considerations show that we are facing a serious problem of ICO. It is true that human beings have adapted to the evolution of technology and the contexts of which they belong. But it is also certain that human capabilities are limited. Therefore, it is important to address ICO on the smart cities subject.

The search for scientific literature that approaches the issue of ICO in smart cities produces very scarce results. For example, Boman (Boman 2012) identifies citizen behavior as a challenge in smart cities, arguing that ICO may occur because *"human capabilities of absorbing information have not increased with increased data flows"*. Boman also identifies the opportunity of using techniques of collaborative information filtering to obtain valuable information and thus overcome the problems associated to ICO. Others explore the eventual usefulness of an unplugged smart city as a model of urban governance, and also as a way of dealing with the information overload issue (Calzada and Cobo 2015). Avoiding technology in some situations, such as on vacation, is also approached by Gretzel *et al.* (2015). Their focus is on smart tourism, which they acknowledge as being closely related with *"extreme technology-dependence"*. They argue that this type of dependence is associated with other issues, like information overload during tourism experiences.

We found that the literature does not frequently address the issue of ICO and, when it does, it proposes possible solutions, such as collaborative filtering. Another aspect is the implicit recognition that smart cities are presented to citizens with a density of technologies which may be disturbing rather than facilitating. For this reason, it is interesting to identify the recognition of *"human capabilities of absorbing information have not increased with increased data flows*" (Boman 2012) or the suggestion of creating opportunities to move away from it (Calzada and Cobo 2015; Gretzel et al. 2015).

3   Information and Communication Overload

Information overload has been identified for several centuries and is an ever-present phenomenon that has renewed challenges over time (Batista and Marques 2017). Information overload mainly affects individuals and organizations, but also carries societal consequences. Today, in the so-called Digital Age, and within smart cities, the issue of information overload has evolved at the same pace as the technological advances (Eppler and Mengis 2004; Melinat et al. 2014). The huge availability of information and the ease of communication exacerbate the worsening of the phenomenon of communication overload and, consequently, of information overload but, on the other hand, offer new ways of dealing with the phenomenon.

Eppler and Mengis (2004) define information overload as a simple notion of dealing with the excess of information. In a decision-making context, information processing and the quality of decision-making increase with the amount of information made available, but just until a given level. When this level is reached, the load of information exceeds one's ability to process it, and thus the phenomenon of information overload is experienced, which negatively contributes to the quality of the decision making (Eppler and Mengis 2004).

Time is another dimension of the information overload issue, proposed by Galbraith (1974) and later reinforced by Tushman and Nadler (1978) among many others. It compares the amount of information that can be integrated into the decision-making process over a period of time with the amount of information that must be included to complete a given process during the same period. In addition to processing capacity and time dimensions, the attribute of information should not be overlooked (e.g. ambiguity, complexity, format, and readability) (Eppler 2015). Other authors state that feelings of stress, anxiety, or confusion may be indicators of ICO (Mulder et al. 2006).

In order to reduce the chance for ICO to occur, it is important to recognize the symptoms, and then to investigate possible causes and solutions. Eppler and Mengis (2004) represent the relationship between causes, symptoms and solutions in a cyclical way, in which the solutions implemented to alleviate the symptoms of informational overload or to eliminate its causes, influence and contribute to new causes of information overload. Thus, there is no solution that permanently solves information overload, but otherwise we are in a permanent cycle of renewal of challenges, which constantly presents new ways of provoking overload.

The effects of ICO are varied. At the individual level, it may manifest with emotional or motivational symptoms (Bawden and Robinson 2009). At the organizational level, it may affect performance and efficiency (Ellwart et al. 2015; Klausegger et al. 2007). Finally, at the societal level, we can observe the overload in the influence provided by the multiple sharing of values, rules and patterns of behavior by the different actors in society (Batista and Marques 2017).

One of the main causes of ICO is the actor involved in the processing - individual, organization or society. ICO may be influenced by the individual's motivation, qualifications, and experience, or by external factors that influence his or her state, such as noise, temperature and time of day. Other causes include the inability to filter and prioritize information, inefficient management of time and the inability to maximize the technological functions at their disposal (Eppler and Mengis 2004; Haase et al. 2015; Ruff 2002). Organizations may cause information overload, because they are dynamic systems. Other causes may be the organizational culture, as cooperative work may lead to situations of ICO due to the increase in communication, as well as the lack of standard procedures or strategies of internal communication (Eppler and Mengis 2004; Reinke and Chamorro-Premuzic 2014; Ruff 2002).

Technology may be another cause of ICO because it is through it that much of information is generated, managed and disseminated. Technology may be an ICO enhancer: very complex or poorly conceptualized IS, low usability, the variety of media of dissemination of information, and its misuse, and weak integration of technologies (Eppler and Mengis 2004; Fuglseth and Sørebø 2014; Harris et al. 2015; Ruff 2002). Complex processes may also influence ICO, since they require the use of great amounts of information. Additionally, the constant interruptions of processes caused, for example, by the simultaneous execution of multiple tasks, the delays of interdependent operations, and the constant requests due to an inadequate communication strategy may also cause ICO (Eppler and Mengis 2004; Jackson and van den Hooff 2012; Lee et al. 2016; Ruff 2002).

4 Critical Factors of Information and Communication Overload in Smart Cities

Intentionality seems to be a rather important factor in processes of developing smart cities. In cases of low intentionality, a number of initiatives by different entities may happen, not necessarily coordinated or sharing common objectives. The diversity of technologies used may be quite varied and poorly integrated, which leads to difficulties in establishing common policies and procedures that contribute to address issues such as ICO. However, other smart cities result from urban management initiatives accordingly to more intentional processes. In this case, it would be easier to establish policies and procedures to reduce the impact of various issues, such as the ICO.

We propose a set of 15 measures that seem to be relevant to take into account situations of ICO in smart cities. These measures constitute a way of observing the issue of ICO and of pointing out ways of action and also of research on smart cities, and thus they are presented here in a generic, however objective, way. Their implementation depends on the possibility of being taken into consideration and implemented in the context of smart cities development initiatives. These measures are a preliminary proposal that needs to be further discussed and validated.

Measures to be taken into consideration in the **identification** of situations of ICO in smart cities: (1) creation of a set of indicators that enhance the identification of possible ICO situations, or situations that drive such overload. The indicators should cover all types of procedures, technology and actors (individual, organizational and societal) involved in smart cities that could be ICO enablers; (2) conceptualization mechanisms for the determination and calculation of the indicators referred to in point (1); (3) monitoring, preferably continuous, of the indicators and their evaluation, in order to activate proactive and possibly reactive mechanisms in ICO situations.

Measures to be taken into consideration in the **prevention** of situations of ICO in smart cities: (4) creation, for each smart city agent, of a list of requirements regarding access, production and dissemination of information and their respective means of processing and communication; (5) definition of appropriate means of access to shared information, as well as efficient integration and coordination of information and their flows by the various smart cities agents, according to the list of requirements mentioned in point (4); (6) definition of a communication policy among the various smart city agents; (7) periodic reassessment of the information requirements of the various smart city agents so that the list identified in point (4) is updated according to the real and current information needs; (8) periodic reassessment of the existing communication policies in order to avoid redundant flows of information and communication among agents; (9) conduct of usability studies of the ICT used in the smart cities, when they are implemented, and verification of their effectiveness for the various actors and contexts; (10) definition of standards for production, formatting and dissemination of data and information, to increase their value and facilitate their integration into the smart city.

Aspects to take into consideration in the creation of possible **solutions** for situations of ICO in smart cities: (11) if the cause of ICO is associated to the amount of information: redefine privileges of access, production and dissemination of information; (12) if the cause of ICO is associated to an excessive or inadequate communication: redefine communication policy; (13) if the cause of ICO is associated to the inadequacy of the technology or of the processes felt by any of the actors or in any of the contexts of use: reassess the usability and the functional requirements of the technologies used and of the processes; (14) if the cause of ICO is associated to the insufficient capacity of processing data and of available information: assess the need to implement other technologies or to resize the existing ones; (15) If the cause of ICO is associated to the attributes and to the value of data and information: adjust the rules of production, formatting and dissemination of data and information.

5   Final Remarks and Conclusion

Both smart cities and ICO have individual, organizational and societal impacts. Other aspects of smart cities that may influence ICO can be mentioned, namely the integration of technologies, the flow of information among multiple smart city's agents, the participating individuals, the intervening organizations, the communities in which they are inserted, the quality of information produced, and the need for communication. These aspects, however, may also be part of the solution to this issue. Hence, it is very relevant to consider ICO in the development and maintenance of smart cities. The measures for

ICO identification, prevention and correction in smart cities presented in this paper are a starting point for further developments in researching this topic.

References

Batista J, Marques RP (2017) An overview on information and communication overload. In: Marques RP, Batista J (eds) Information and communication overload in the digital age. IGI Global, Hershey, PA, USA, pp 1–19. doi:10.4018/978-1-5225-2061-0.ch001

Bauman Z (2013) Liquid times: living in an age of uncertainty. Wiley, Hoboken

Bawden D, Robinson L (2009) The dark side of information: overload, anxiety and other paradoxes and pathologies. J Inf Sci 35:180–191

Boman M (2012) Digital Cities EIT ICT Labs:12

Calzada I, Cobo C (2015) Unplugging: deconstructing the smart city. J Urban Technol 22:23–43

Ellwart T, Happ C, Gurtner A, Rack O (2015) Managing information overload in virtual teams: effects of a structured online team adaptation on cognition and performance. Eur J Work Organ Psychol 24:812–826

Eppler MJ (2015) Information quality and information overload: the promises and perils of the information age Communication and Technology, vol. 5, p. 215

Eppler MJ, Mengis J (2004) The concept of information overload: A review of literature from organization science, accounting, marketing, MIS, and related disciplines. Inf Soc 20:325–344

Fuglseth AM, Sørebø Ø (2014) The effects of technostress within the context of employee use of ICT. Comput Hum Behav 40:161–170

Galbraith JR (1974) Organization design: an information processing view. Interfaces 4:28–36

Giovannella C (2014) Smart learning eco-systems: "fashion" or "beef"? J e-Learn Knowl Soc 10

Gretzel U, Sigala M, Xiang Z, Koo C (2015) Smart tourism: foundations and developments. Electron Markets 25:179–188

Haase RF, Ferreira JA, Fernandes RI, Santos EJ, Jome LM (2015) Development and validation of a revised measure of individual capacities for tolerating information overload in occupational settings. J Career Assess: 1069072714565615

Harris KJ, Harris RB, Carlson JR, Carlson DS (2015) Resource loss from technology overload and its impact on work-family conflict: can leaders help? Comput Hum Behav 50:411–417

Jackson TW, van den Hooff B (2012) Understanding the factors that effect information overload and miscommunication within the workplace. J Emerg Trends Comput Inf Sci 3:1240–1252

Klausegger C, Sinkovics RR, "Joy" Zou H (2007) Information overload: a cross-national investigation of influence factors and effects. Marketing Intell Plan 25:691–718

Lee AR, Son S-M, Kim KK (2016) Information and communication technology overload and social networking service fatigue: a stress perspective. Comput Hum Behav 55:51–61

Maslow AH (1943) A theory of human motivation. Psychol Rev 50:370

Matsuda K (2016) Hyper-reality. https://www.youtube.com/watch?v=YJg02ivYzSs&t=64s. Accessed March 2017

Melinat P, Kreuzkam T, Stamer D (2014) Information overload: a systematic literature review. In: Johansson B, Andersson B, Holmberg N (eds) Perspectives in business informatics research: 13th international conference, BIR 2014, Lund, Sweden, 22–24 September, 2014. Proceedings. Springer International Publishing, Cham, pp 72–86

Mulder I, de Poot H, Verwij C, Janssen R, Bijlsma M (2006) An information overload study: using design methods for understanding. In: Proceedings of the 18th Australia conference on computer-human interaction: design: activities, artefacts and environments, pp 245–252. ACM

Neirotti P, De Marco A, Cagliano AC, Mangano G, Scorrano F (2014) Current trends in Smart City initiatives: some stylised facts. Cities 38:25–36

Reinke K, Chamorro-Premuzic T (2014) When email use gets out of control: Understanding the relationship between personality and email overload and their impact on burnout and work engagement. Comput Hum Behav 36:502–509

Ruff J (2002) Information overload: causes, symptoms and solutions Harvard Graduate School of Education, pp 1–13

Tushman ML, Nadler DA (1978) Information processing as an integrating concept in organizational design. Acad Manag Rev 3:613–624

Designing for Sustainable Urban Mobility Behaviour: A Systematic Review of the Literature

Lisa Klecha[(⊠)] and Francesco Gianni

Department of Computer Science, Norwegian University of Science and Technology, Trondheim, Norway
lisark@stud.ntnu.no, francesco.gianni@ntnu.no

Abstract. Urban mobility is a challenge in cities undergoing growing urbanisation, requiring a shift in behaviour towards more sustainable means of transportation. To investigate how technology can mediate the process of behaviour change, particularly in the context of smart cities, this paper presents a systematic literature review. Three areas are of interest: the utilised technology, behaviour change strategies, and citizen participation in the development process. A total of 14 different applications were included in the final review. The findings show that mobile devices are being prevalently used, persuasive strategies are foremost mentioned, and end-user involvement is happening late in the development process, serving primarily testing purposes. This points out that there are still various unexplored possibilities. It is suggested that future research should explore opportunities stemming from ubiquitous technology, employ behaviour change strategies grounded in reflective learning, and promote citizen involvement with participatory methods.

Keywords: Sustainable mobility · Behaviour change · Smart city · Reflection

1 Introduction

Cities are increasingly gaining in importance and are considered driving forces of the future (Albino et al. 2015). This development stems foremost from growing urbanisation, with more people moving and living in urban areas. As of 2016, 54.5% of the world's population were estimated to be living in urban settlements, current trends forecast an increase to 60% in 2030 (United Nations and Social Affairs 2016). Rapid growth assigns cities a significant social, economic and environmental function (Albino et al. 2015), but also lets them face complex challenges in the process. In this context, the concept of "smart cities" has been coined to indicate cities that devise strategies to mitigate those challenges in a smart way (Chourabi et al. 2012). Some of these challenges are indicated by greenhouse gas emissions. According to Hildermeier and Villareal (2014), urban transport makes up approximately a quarter of CO_2 emissions with respect to overall transport. With 73.7% cars represent thereby the prevalent form of travel. Urban transport hence exerts ample influence on the quality of life in a city, being a causer of traffic congestion, noise- and air pollution, having harmful effects on

© Springer International Publishing AG 2018
Ó. Mealha et al. (eds.), *Citizen, Territory and Technologies: Smart Learning Contexts and Practices*,
Smart Innovation, Systems and Technologies 80, DOI 10.1007/978-3-319-61322-2_14

public health. Consequently, cities need to respond to this issue by reducing car usage and increase the use of more sustainable means of transport, such as walking, biking or public transportation, making a change in behaviour inevitable.

Technology is noted for being a "key driver" within smart city initiatives (Chourabi et al. 2012). It can also be utilised to foster behaviour change, here the notion of "persuasive technology" is well-established. Persuasive technology, that is "any interactive computing system designed to change people's attitudes, behaviours or both" (Pettersen and Boks 2008; Fogg 2003, p. 288). The idea of persuasive technology has certainly undergone criticism (Atkinson 2006), and some of its strategies have been contested regarding their ethics (Pettersen and Boks 2008), and long-term applicability, with Brynjarsdottir et al. (2012, p. 951) claiming that persuasive sustainability system's "long-term success is susceptible to being undermined by factors outside of what it aims to measure and control". Considering this, merely relying on persuasive strategies, guided by prescribed goals, may not result in durable behaviour change, nor does it appear to be suitable in a smart city context, in which projects focus on "more informed, educated, and participatory citizens" (Chourabi et al. 2012).

An alternative approach is that of reflective learning, a mechanism emphasising self-directed learning through experience. Krogstie et al. (2013) introduced a reflective learning cycle, short CSRL model, which offers the possibility of individual as well as collaborative learning. The overall model is composed of four main stages, namely plan and do work (1), initiate reflection (2), conduct reflection session (3) and apply outcome (4), with diverse activities at each stage. Notable aspects of the model are, that it considers the absence of teachers, sees reflective learning as a highly iterative process, and acknowledges the importance of social aspects in reflection, as it is frequently accomplished collaboratively. These characteristics may make it hence a suitable model to be employed in a smart city setting.

Still, particularly pertaining to the issue of ethics in behaviour change applications, active citizen involvement needs to be ensured when developing such interventions. According to Pettersen and Broks (2008), design methods, such as user-centred and participatory design, offer numerous advantages for the development of a service. The authors further state that, technology that is developed democratically is more likely to meet citizen needs, and ensures that the user's freedom is maintained, with persuasive technologies aiming at steering individual behaviour towards goals that are not set by the users themselves. Furthermore, participation and empowerment can instil a feeling of ownership in citizens. Eventually, such an undertaking requires effective means of engaging the public and stakeholders in co-design.

2   Motivation and Research Questions

To highlight the contribution of this review, preliminary searches were undertaken to identify similar reviews. A review on persuasive technologies for sustainable urban mobility was published by in April of 2016 (Anagnostopoulou et al. 2016). While the theme of "urban mobility behaviour change applications" remains the same, the review at hand emphasises other facets in the studies. Anagnostopoulou et al.'s review evolves around the notion of persuasive technology, and while the review states that user characteristics and context should be considered, users have not been the unit of analysis with respect to overall involvement. The review at hand contributes to the theme by focusing on citizen participation and behaviour change through reflection, with the two approaches seen as holding great opportunities in a smart city context. This effort is summarised in the research questions below:

RQ1:   Which technologies have been utilised for behaviour change applications?
RQ2:   Which behaviour change strategies have been employed?
RQ3:   Have the final users of the application been involved in the application development? If so, when, to what degree and with what methods?

The paper is then organised as follows: In Sect. 3, the process used for the literature review is described. Section 4 presents the results of the review, in accordance with the research questions. Section 5, discusses then findings and implications. Finally, Sect. 6 summarises the results and provides perspectives for research.

3   Review Method and Process

A systematic literature review facilitates the process of summarising existing research, assessing where there are gaps, and allows for new research to be placed within the existing research base. This method was chosen, to gain an overview of the issue of urban mobility, and on how this matter had been addressed by applications previously. The search strategy was thereby informed by the guidelines for systematic literature reviews, described by Kitchenham and Charters (2007).

Data Sources
Relevant documents for the search were retrieved, using two approaches: Foremost through keyword based searches of online databases, and secondly manual screening of conference proceedings of former Persuasive Technology conferences (http://persuasivetechnology.eu/), as well as the dblp computer science bibliography (http://dblp.uni-trier.de/). Web of Science (https://apps.webofknowledge.com/).

ACM digital library (https://dl.acm.org/), Elsevier ScienceDirect (https://www.sciencedirect.com/), Scopus (http://www.scopus.com/), and Springer (http://rd.springer.com/) were chosen as online databases and queried using keywords, assembled into search strings. Those databases were chosen due to their eligibility for the topic, and because they allowed a large number of search terms.

Search and Keywords
As suggested by Kitchenham and Charters (2007), and applied by Gianni and Divitini (2016) in a similar setting, the PICOC framework was utilised to guide the selection of the keywords. PICOC is an abbreviation, constituting the words population, intervention, comparison, and context, which classify the keywords accordingly. Initial keywords were synthesised from the research questions, and expanded through respective synonyms and recombinations, that emerged during iterative pilot searches. Search terms pertaining to ICT were omitted, to lower the amount of constraints, and rather assess manually the suitability of the retrieved documents.

Table 1. Keywords for the search string

Population	–
Intervention	Sustainable mobility, sustainable travel, sustainable transport, sustainable transportation, green mobility, green travel, green transport, green transportation, personal mobility, personal travel, personal transport, personal transportation
Comparison	–
Outcome	Behaviour change, behavioural change, behaviour choice, behaviour promotion, behaviour encouragement, behaviour management, mobility behaviour, travel behaviour, transport behaviour, transportation behaviour, habit
Context	City, cities, smart city, urban, connected city, intelligent city, digital city

Table 1 lists the chosen keywords in the PICOC structure. The terms within each row were connected through Boolean OR operators, and the three individual rows were linked using Boolean AND operators. In most databases, it was possible to choose the abstract as search field. Table 2 summarises the outcomes of the searches. The field "Topic" combines here the title, abstract, keyword and index fields of the article. In addition, manual searches retrieved 3 studies.

Table 2. Results from the online databases without duplicates removed

Database	Documents	Field	Date of Search
Web of science	137	Topic	11.11.2016
ACM	7	Abstract	11.11.2016
Science direct	41	Abstract	11.11.2016
Scopus	136	Abstract	11.11.2016
Springer	17	Overall	11.11.2016
Total	338		

Screening of Papers

Following the search, the metadata of all documents was organised in a spreadsheet. This facilitated the ordering of the documents by title, and the subsequent removal of duplicates. The remaining documents were then screened individually by the author and supervisor. Inclusion or exclusion decisions were based on the title and the abstract, and guided by the criteria outlined below. This approach yielded in 29 potentially relevant documents. During a second collaborative screening, this number was reduced to a total of 15 documents that were included in the final review, discussing 14 different applications.

Report Eligibility. A report had to be in English; published on a database, conference or journal that is scientific; published in or after 2005 (the notion of "smart city" gained notoriety between 2005 and 2007 (Gianni and Divitini 2016)); fully accessible; short or long, posters and demos had to be excluded.

Study Eligibility. A study had to address behaviour change of people; outline technology in the abstract; solely address sustainable mobility; be a primary source.

4 Results and Findings

A brief summary of all reviewed studies is given in Table 3.

Table 3. Summary of reviewed studies

Reference	Summary
(Gotzenbrucker and Kohl 2012)	AnachB is an advanced traveller information system, enabling citizens to compare routes and different mobility modes in real-time, and provides users with information all around planned trips.
(Magliocchetti et al. 2012)	i-Tour, an ambient intelligence system, is as a travel mobility assistant in a multi-modal setting. The service employs context-awareness, inferring information about the user and the environment.
(Gabrielli and Maimone 2013)	Superhub is a mobile system utilising diverse motivational strategies to influence citizens' mobility choices. A local journey planning service is provided.
(Bothos et al. 2012, 2013; Bothos et al. 2014)	The application Peacox is discussed at different stages in (2012), specifying technological details on how routes are recommended, (2013), informing about the use of choice architecture for a travel recommender, and (2014) placing emphasis on persuasive strategies. Peacox provides users with route options suiting their preferences and contexts, and sorts these according to their CO_2 impact.
(Gabrielli et al. 2014)	MatkaHupi, automatically tracks journeys and emissions and fosters sustainable mobility choices through challenges. This paper furthermore introduces a journey planner system which employs a journey diary.
(Cheng and Lee 2015)	BikeTogether, a mobile app for cycle commuting, allows users to figuratively cycle together while being connected over the Internet. A bicycle flashlight is used as a metaphor for users to feel accompanied and guided.
(Semanjski et al. 2016)	Describes a web portal and a mobile application. While the web portal allows users to specify personal, usual travel information, the mobile application senses mobility behaviour and provides the user with travel suggestions, favouring sustainable options, and offers self-monitoring and social activities.

(*continued*)

Table 3. (*continued*)

Reference	Summary
(Monzon et al. 2013)	A Real-Time Passenger Information System installed in buses and stops or interchange stations. Multi-modal information regarding, departure, connections or travel time is visualised on screens and on a webpage.
(Wernbacher et al. 2015)	Traces employs elements from gamification, serious games, and pervasive games, to promote the use of sustainable mobility. The mobile application features both city related quizzes and offline quests. The game goal is to leave colourful digital traces on a city map through the use of multi-modal mobility.
(Broll et al. 2012)	Tripzoom is a concept comprising a web portal, a mobile application, and a city dashboard. The web portal is linked to social networks, the mobile application supports users in understanding their mobility behaviour, and the city dashboard lets incentives and rewards be managed by individual cities.
(Kazhamiakin et al. 2015)	Viaggia Rovereto, a mobile route planning application, incentivises sustainable choices through gamification. Sustainable route options are highlighted, and users are rewarded when traveling sustainably.
(Bordin et al. 2014)	ViaggiaTrento is a multi-modal trip planning application, developed utilising participatory design practices with students, and incorporating collaborative efforts from its users.
(Wunsch et al. 2015)	Discusses persuasive strategies, encouraging bike usage, such as triggering messages, or a virtual bike tutorial aiming at increasing biker's self-efficacy towards biking.

Technology utilised for behaviour change applications

Technological solutions, harnessed to support the respective applications, are visualised in Fig. 1. Smartphones emerge here as the prevailing theme, justified with, among others, their widespread use and their ability to function as sensors, thus

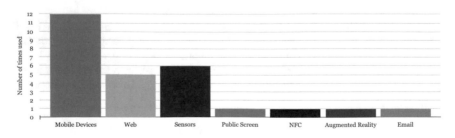

Fig. 1. Technology used for the behaviour change interventions

obtaining information from the environment, shedding light on a user's behaviour (Semanjski et al. 2016; Broll et al. 2012).

Prevalently, mobile applications feature either a route planning system (Gabrielli and Maimone 2013; Gotzenbrucker and Kohl 2012; Magliocchetti et al. 2012; Bothos et al. 2013; Semanjski et al. 2016; Kazhamiakin et al. 2015; Bordin et al. 2014; Bothos et al. 2014; Gabrielli et al. 2014) or are solely focused on illustrating people's mobility behaviour (Cheng and Lee 2015; Wernbacher et al. 2015; Broll et al. 2012; Wunsch et al. 2015). Sensors embedded in smartphones, provide data that is used to infer information about a user's movement (Magliocchetti et al. 2012; Bothos et al. 2013; Cheng and Lee 2015; Semanjski et al. 2016; Broll et al. 2012; Bothos et al. 2014; Gabrielli et al. 2014) (e.g. standing, walking, cycling, use of public transport) and ambient conditions (Magliocchetti et al. 2012) (e.g. brightness, noise). Sensors are also present in the environment (Magliocchetti et al. 2012; Broll et al. 2012; Bordin et al. 2014), collecting information, used in the applications. Web applications provide real-time public transport information (Monzon et al. 2013), support users in their mobility decisions with a route planner (Gotzenbrucker and Kohl 2012), or collect and visualise mobility behaviour data of a user (Semanjski et al. 2016; Wunsch et al. 2015) or a community of users (Broll et al. 2012).

Strategies employed to guide behaviour change

The review comprises both applications, specifically mentioning persuasive technology or strategies as their approach, and interventions that do not state a methodology. In both cases the applications were categorised according to their usage of one, or several persuasive strategies. Krogstie et al.'s (2013) CSRL model provided the theoretical underpinning for the identification and assessment of reflection as the employed behaviour change methodology. In the following, the foremost used methods are outlined and complemented by one example of use in the applications.

Persuasive Technology. Oinas-Kukkonen and Harjumaa (2008) categorise persuasive techniques within four categories: primary task, dialogue, system credibility, and social support. Each category outlines several design principles, which have been mapped onto the reviewed studies.

Within the primary task support category Reduction, Tunneling and Self-Monitoring were found to be most frequently employed. Reduction manifests itself in the provision of journey planning services, easing the accessibility of mobility information, for instance by sorting route suggestions according to their environmental friendliness (Magliocchetti et al. 2012; Semanjski et al. 2016; Kazhamiakin et al. 2015). Tunneling is afforded by sustainability challenges or quests (Gabrielli et al. 2014; Wernbacher et al. 2015; Broll et al. 2012), and Self-Monitoring is enabled through the provision of graphical, statistical representations of reported or logged behaviour, comprising, for instance, modes of transport used (Gabrielli and Maimone 2013; Gabrielli et al. 2014; Semanjski et al. 2016; Wernbacher et al. 2015; Broll et al. 2012; Wunsch et al. 2015).

In the dialogue support category, Rewards are the most commonly applied strategy, awarded to users which opt for sustainable mobility choices (Magliocchetti et al. 2012;

Gabrielli and Maimone 2013; Wernbacher et al. 2015; Kazhamiakin et al. 2015). However, none of the systems outlined any of the principles of the credibility support category clearly. Social Facilitation is the prevalently applied strategy within the social support category, followed by Social Comparison. Social Facilitation is supported by a shared view on a leader board, listing users performing the behaviour (Gabrielli and Maimone 2013; Gabrielli et al. 2014; Semanjski et al. 2016). Also, Social Comparison is supported by leader boards, changing images (Broll et al. 2012) or colours (Cheng and Lee 2015), comparing users.

Reflective Learning. None of the reviewed studies names reflection as the core principle guiding behaviour change. Nevertheless, some of the approaches, found in the studies at hand, have the capability of supporting the CSRL model's four stages: plan and do work (1), initiate reflection (2), conduct reflection session (3) and apply outcome (4). The studies were thus mapped onto the stage specific activities, to observe to what degree existing applications are potentially able to facilitate reflective learning. All presented applications support the work, when being viewed as the "adoption of sustainable transportation means".

"Plan work" and "do work" are primarily supported by applications providing a journey planning service, allowing users to plan and conduct their commute. The "plan work" activity, is furthermore supported by allowing users to set personalised, sustainable goals they wish to accomplish (Gabrielli and Maimone 2013; Bothos et al. 2013; Gabrielli et al. 2014). The work is monitored by either users manually reporting their behaviour (Magliocchetti et al. 2012; Gabrielli et al. 2014, 2014; Kazhamiakin et al. 2015; Wunsch et al. 2015; Wernbacher et al. 2015), or by the system logging behaviour automatically (Magliocchetti et al. 2012; Bothos et al. 2013; Gabrielli et al. 2014; Cheng and Lee 2015; Semanjski et al. 2016; Broll et al. 2012; Bothos et al. 2014).

Three ways can then be identified, in which the applications initiate reflection and set a reflection objective. Firstly, CO_2 emissions being displayed along with suggested routes, make users aware of their potential actions (Magliocchetti et al. 2012; Bothos et al. 2013, 2012; Bothos et al. 2014). Secondly, sustainability encouraging messages or reminders, sent to the user, draw attention to sustainable issues and possibilities in behaviour (Gabrielli and Maimone 2013; Bothos et al. 2013; Gabrielli et al. 2014; Cheng and Lee 2015; Wunsch et al. 2015). Thirdly, sustainability challenges may set a particular objective for one's behaviour, that can be reflected upon (Gabrielli et al. 2014; Semanjski et al. 2016; Broll et al. 2012).

The stage concerned with conducting the reflection session is partially supported in some applications by displaying the CO_2 emissions along with the respective routes, making related information available, and by visualisations of behavioural information, reconstructing work experiences.

The final stage of applying the outcome is facilitated in part in (Bothos et al. 2013) by providing the user with feedback on personal, previously caused CO_2 emissions, highlighting the issue requiring change. The app also provides the user with routes that would aid reducing the said emissions.

End-user involvement in the application development

To highlight the respective point of involvement and the degree of participation, the systems development life cycle and typologies of citizen participation were utilised to frame the assessment. The systems development life cycle as described in (Majid et al. 2010) depicts five steps in the development of an information system: (1) project selection and planning, (2) requirement analysis, (3) system design, (4) development, and (5) testing and deployment. Pertaining to the overall distribution, there appears to be a significant user involvement during the testing phase, primarily constituting itself through the implementation of user studies (Gotzenbrucker and Kohl 2012; Gabrielli and Maimone 2013; Gabrielli et al. 2014; Cheng and Lee 2015; Semanjski et al. 2016; Monzon et al. 2013; Kazhamiakin et al. 2015; Bordin et al. 2014; Bothos et al. 2014; Wunsch et al. 2015). Involvement in the other stages is rarely implemented. Eventually, (Bordin et al. 2014) poses as the only reviewed article involving users at all stages through social innovation.

Capra (2014), in the context of smart city citizen participation, merged traditional typologies of citizen participation, and concepts of social innovation. Citizen participation, as defined by Capra, reaches from "information", a mono-directional flow of information, to "social innovation", being citizen led, collective initiatives that address their specific needs. Those typologies built the basis for the assessment of degrees of citizen participation and participatory methods utilised in the reviewed studies. Figure 2 illustrates which types of participation foremost took place, whereas Fig. 3 visualises the methods used to facilitate participation.

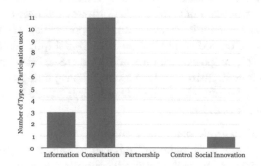

Fig. 2. Types of Participation, categories taken from (Capra Capra 2014)

Users foremost contribute through consultation, a bi-directional flow of information intended to gather feedback. This involvement constitutes itself primarily through questionnaires, allowing to observe changes in, for instance, participants travel behaviour (Gotzenbrucker and Kohl 2012; Gabrielli and Maimone 2013; Gabrielli et al. 2014). Furthermore interviews (Gotzenbrucker and Kohl 2012; Gabrielli and Maimone 2013; Gabrielli et al. 2014; Bothos et al. 2014) and diaries (Gotzenbrucker and Kohl 2012) give information about the user's experience with a system. Focus groups are used to collect feedback on an ideal design (Gotzenbrucker and Kohl 2012), issues with current conditions (Bordin et al. 2014), application usage and behavioural changes (Bothos et al. 2014).

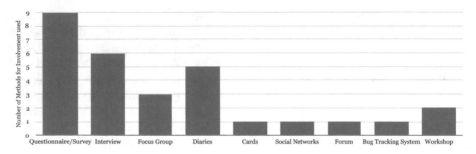

Fig. 3. Methods used for User/Citizen Involvement

5   Discussion

Technologies utilised for behaviour change applications

Mobile devices emerge as prevalently used technical mediators, with 12 reviewed applications envisioning or implementing services being supported by them.

A systematic mapping of technology-enhanced smart city learning, conducted by Gianni and Divitini (Gianni and Divitini 2016), revealed similar results, identifying unexplored technical opportunities with interactive objects and the Internet of Things (IoT). In a later article the authors then also pinpointed limitations of smartphone apps, when being used for learning in smart cities, including that of sustainable behaviours (Gianni et al. 2016). The essential shortcomings are the restricted interaction strategies provided by mobile devices, hindering the tailoring of the user experience towards a specific scenario. Here ubiquitous computing opportunities such as tangible user interfaces, affording an embodied interaction with digital information, and augmented objects are alternatively proposed, with the ability to create "rich and unobtrusive user experiences" (Gianni et al. 2016, p. 13), allowing the capturing of various data types. Therefore, besides the benefits of mobile technologies, a large body of yet unexplored technological possibilities exists.

Strategies employed to guide behaviour change

Reduction, tailoring, and self-monitoring emerge as prevalent persuasive strategies. These, and other persuasive strategies or application functionalities have been found to support aspects of the CSRL model. While none of the studies outlines reflection as guiding behaviour change, some report self-reflection, as being supported by application features (Gabrielli and Maimone 2013; Gabrielli et al. 2014; Wernbacher et al. 2015), or as subject for future research (Bothos et al. 2014). Given earlier outlined criticism of persuasion, reflection may represent as an alternative approach. This is supported by Brynjarsdottir et al. (2012, p. 954), proposing that behaviour should not merely be prescribed, rather, systems should foster open-ended reflection, meaning for users to "reflect on what it actually means to be sustainable in a way that makes sense in the context of their own lives".

End-user involvement in the application development

There appears to be a significant user involvement during the testing phase, primarily constituting itself through the implementation of user studies. Involvement in the other stages is rarely implemented. Eventually, (Bordin et al. 2014) is the only reviewed article involving users at all stages.

During user studies, users primarily contribute through consultation. While focus groups or interviews may allow participants to voice their opinions more freely, questionnaires and similar approaches are potentially more restrictive. Moreover, how feedback is eventually exploited and translated into system functionality might not be up to the participants to decide. Also, Brynjarsdottir et al. (2012) found, that desirable behaviour is foremost predetermined, in a top-down approach, by professionals involved in the development of behaviour change systems, with only 3 of their 36 reviewed papers stating participatory design. Eventually, the authors suggest including users more thoroughly in the design process through participatory design methods, with the prospect of ensuring that the system's definition of sustainability corresponds to user's daily life, thus making the applications more accepted and useful. This approach is also supported by Pettersen and Broks (2008) to democratise application development and empower people. In terms of the IoT, tools such as TILES Toolkit (http://tilestoolkit.io/) can be utilised to foster end-user participation, allowing non-experts to design and prototype smart objects.

6 Conclusion

Three themes have been guiding this review, namely the utilised technology, the applied behaviour change strategies and the degree end-user participation in the application development. Altogether the outcomes imply that various possibilities in the above areas are still unexplored, which is underpinned by previous research in related areas. Future research efforts will focus on exploring ways in which end-users can actively and meaningful participate in the stages of design and development. Ultimately, an ideal solution would create conditions that empower and inform citizens, enabling them to create their own change through social innovation.

References

Albino V, Berardi U, Dangelico RM (2015) Smart cities: definitions, dimensions, performance, and initiatives. J Urban Technol 22(1):3–21

Anagnostopoulou E, Bothos E, Magoutas B, Schrammel J, Mentzas G (2016) Persuasive technologies for sustainable urban mobility. *arXiv Preprint* arXiv:1604.05957

Atkinson BMC (2006) Captology: a critical review. In: International conference on persuasive technology, pp 171–82. Springer

Bordin S, Menendez M, De Angeli A (2014) ViaggiaTrento: an application for collaborative sustainable mobility. ICST Trans Ambient Syst 4:e5

Bothos E, Dimitris A, Gregoris M (2012) Recommending eco-friendly route plans. In: Proceedings of 1st international workshop on recommendation technologies for lifestyle change, pp 12–17

Bothos E, Dimitris A, Gregoris M (2013) Choice architecture for environmentally sustainable urban mobility. In: CHI 2013 Extended abstracts on human factors in computing systems, pp 1503–1508. ACM

Bothos E, Prost S, Schrammel J, Röderer K, Mentzas G (2014) Watch your emissions: persuasive strategies and choice architecture for sustainable decisions in urban mobility. PsychNology J 12(3):107–126

Broll G, Cao H, Ebben P, Holleis P, Jacobs K, Koolwaaij J, Luther M, Souville B (2012) Tripzoom: an app to improve your mobility behavior. In: Proceedings of the 11th international conference on mobile and ubiquitous multimedia, p. 57. ACM

Brynjarsdottir H, Håkansson M, Pierce J, Baumer E, DiSalvo C, Sengers P (2012) Sustainably unpersuaded: how persuasion narrows our vision of sustainability. In: Proceedings of the SIGCHI conference on human factors in computing systems, pp 947–956. ACM

Capra CF (2014) The Smart City and Its Citizens: Governance and Citizen Participation in Amsterdam Smart City. Mastersthesis, Erasmus University Rotterdam

Cheng Y-M, Lee C-L (2015) Persuasive and engaging design of a smartphone app for cycle commuting. mUX J Mobile User Exp 4(1):1

Chourabi H, Nam T, Walker S, Gil-Garcia JR, Mellouli S, Nahon K, Pardo TA, Scholl HJ (2012) Understanding smart cities: an integrative framework. In: 2012 45th hawaii international conference on system science (HICSS), pp 2289–2297. IEEE

Fogg, BJ (2003) Persuasive technology: using computers to change what we think and do. Interactive technologies series. Morgan Kaufmann Publishers. https://books.google.no/books?id=9nZHbxULMwgC

Gabrielli S, Forbes P, Jylhä A, Wells S, Sirén M, Hemminki S, Nurmi P, Maimone R, Masthoff J, Jacucci G (2014) Design challenges in motivating change for sustainable urban mobility. Comput Hum Behav 41:416–423

Gabrielli S, Maimone R (2013) Are change strategies affecting users' transportation choices? In: Proceedings of the biannual conference of the Italian chapter of SIGCHI, p 9. ACM

Gianni F, Divitini M (2016) Technology-enhanced smart city learning: a systematic mapping of the literature. Interact Design Archit (S) J 27:28–43

Gianni F, Mora S, Divitini M (2016) IoT for smart city learning: towards requirements for an authoring tool. In: CEUR workshop proceedings, vol 1602

Gotzenbrucker G, Kohl M (2012) Advanced traveller information systems for intelligent future mobility: the case of'AnachB'in Vienna. IET Intell Transp Syst 6(4):494–501

Hildermeier J, Villareal A (2014) Two ways of defining sustainable mobility: Autolib' and BeMobility. J Environ Policy Plann 16(3):321–336

Kazhamiakin R, Marconi A, Perillo M, Pistore M, Valetto G, Piras L, Avesani F, Perri N (2015) Using gamification to incentivize sustainable urban mobility. In: 2015 IEEE first international on smart cities conference (ISC2), pp 1–6. IEEE

Kitchenham B, Charters S (2007) Guidelines for performing systematic literature reviews in software engineering

Krogstie BR, Prilla M, Pammer V (2013) Understanding and supporting reflective learning processes in the workplace: the CSRL model. In: European conference on technology enhanced learning, pp 151–64. Springer

Magliocchetti D, Gielow M, De Vigili F, Conti G, de Amicis R (2012) Ambient intelligence on personal mobility assistants for sustainable travel choices. JUSPN 4(1):1–7

Majid RA, Noor NLM, Adnan WAW, Mansor S (2010) A survey on user involvement in software development life cycle from practitioner's perspectives. In: 2010 5th international conference on computer sciences and convergence information technology (ICCIT), pp 240–43. IEEE

Monzon A, Hernandez S, Cascajo R (2013) Quality of bus services performance: benefits of real time passenger information systems. Transp Telecommun 14(2):155–166

Oinas-Kukkonen H, Harjumaa M (2008) Towards deeper understanding of persuasion in software and information systems. In: 2008 first international conference on advances in computer-human interaction, pp 200–205. IEEE

Pettersen IN, Boks C (2008) The ethics in balancing control and freedom when engineering solutions for sustainable behaviour. Int J Sustain Eng 1(4):287–297

Semanjski I, Aguirre AJL, De Mol J, Gautama S (2016) Policy 2.0 platform for mobile sensing and incentivized targeted shifts in mobility behavior. Sensors 16(7):1035

United Nations, Department of Economic, and Population Division Social Affairs (2016). The World's Cities in 2016 – Data Booklet (ST/ESA/ SER.A/392)

Wernbacher T, Pfeiffer A, Platzer M, Berger M, Krautsack D (2015) Traces: a pervasive app for changing behavioural patterns. In: European conference on games based learning, vol 589. Academic Conferences International Limited

Wunsch M, Stibe A, Millonig A, Seer S, Dai C, Schechtner K, Chin RCC (2015) What makes you bike? exploring persuasive strategies to encourage low-energy mobility. In: International conference on persuasive technology, pp 53–64. Springer

Public Policies for Quality Assurance in Distance Learning Towards Territory Development

Lurdes Nakala[1(✉)], António Franque[1], and Fernando Ramos[2]

[1] INED - Instituto Nacional de Educação à Distância, Maputo, Mozambique
{lurdesnakala,afranque}@ua.pt
[2] Department of Communication and Art, CIC.Digital/DigiMedia,
University of Aveiro, Aveiro, Portugal
fernando.ramos@ua.pt

Abstract. This paper is about public policies for quality assurance in distance learning towards territory development. The study was based on the literature review which enabled an overview of the key concepts for a better understanding of the theme and documental analysis to bring experiences from Mozambique. Challenges for distance learning pubic policies that recognize different contexts are also presented, as well as suggestions for quality policies conducive to the local development; and a short note for further work is also included.

Keywords: Distance learning · Public policies · Quality assurance · Territory development

1 Introduction

Distance Learning (DL) as a flexible mode can contribute to the territory development through settlement of inhabitants on their own environment, avoiding displacement in search of educational opportunities, which very often are in the cities, as is the case of Mozambique. DL can also lessen the demographic pressure over urban areas from people seeking academic institutions, as well as better socio-economic conditions. Through DL people can acquire contextualized qualifications based on the specific needs of a territory. This paper presents public policies for quality assurance in distance learning towards territory development through literature review, which enabled an overview of the key concepts for a better understanding of the theme, as well as documental analysis to bring experiences from Mozambique. The following section (Sect. 2) is an overview of the concepts on quality assurance, public policies and distance learning; Sect. 3 is about the contribution of DL in the territory development, followed by Sect. 4 that deals with the challenges to quality public policies towards territory development, including future work.

© Springer International Publishing AG 2018
Ó. Mealha et al. (eds.), *Citizen, Territory and Technologies: Smart Learning Contexts and Practices*,
Smart Innovation, Systems and Technologies 80, DOI 10.1007/978-3-319-61322-2_15

2   Quality Assurance, Public Policies and Distance Learning

There is no doubt that discussing quality assurance public policies in DL is extremely important because education plays a huge role in preparing people for life and for the efficiency and effectiveness of all socio-economic sectors. Quality in education is the center of attention at the global level and is a recommendation of the United Nations Sustainable Development Goals Agenda 2015–2030 (UNDP 2015).

The growing concern in quality of education provided has led to different interpretations of what quality is and how should it be maintained, and even increased. Regarding the first issue, quality is perceived as a "philosophical concept" (Green 1994, 27) and "quality is often referred to as a relative concept" (Harvey and Green 1993, 10). Authors believe that quality is everything with perfection, excellence, fitness for a certain purpose, value for money and transformation (Harvey and Green 1993; Green 1994; Harvey 2006). In the context of education, such view may not be appropriate due to the dynamics that occur and involve people interactions, though the concepts may serve as the basis for the understanding of quality assurance mechanisms in educational settings.

As raised by Harvey (2006, 1) "Quality assurance is about checking the quality…". Vlăsceanu et al. (2007) advanced the meaning of quality assurance as an internal process to achieve the vision, mission and objectives of the institution. However, considering quality assurance only from an internal perspective of the institution would be neglecting the government's responsibility for provision of DL programs.

Indeed, quality assurance in DL involves both internal and external processes. Internally, it is an endogenous monitoring mechanism to verify the degree of achievement of the vision, mission and objectives of the institution, to improve the offering of programs. Externally, quality assurance is primarily the responsibility of government or non-governmental regulatory institutions that play an important role in safeguarding the standards of education established through respective public policies.

Based on different perceptions of quality due to the different interest groups, such as government, teachers, students, parents, employers and the community in general, it would be sensible for quality assurance mechanisms to accommodate these different quality focuses. In this sense, depending on the specific context in which public policy is defined, we may find different models, especially in relation to established quality standards, although all models tend to share basic procedures and principles.

With the rapid development of communication technologies, which contributes to the emergence of new approaches to education, leading to enrichment of the development of DL, in the coming years is estimated to account for 30% of the educational supply (Ossiannilsson et al. 2015). Indeed, the development of technologies has driven the evolution of social media networks, which in turn have a great impact on the way of approaching DL practices (Peters 2004; Bates 2008). In teaching and learning in a digital context, the user has the possibility, from self-instructional and other materials, to interact dynamically with the study contents. The student passes from a passive subject, becomes active and participates individually and collectively in the process of knowledge production. The electronic availability and the possibility of combining the various support tools allow the creation of more integrated learning environments, which facilitates distance study. These dynamics require provision of responsive quality public policies, which in DL generally its implementation is carried out through accreditation processes.

In Mozambique, the full process of accreditation involves three main governmental institutions, namely: *Instituto Nacional de Educação à Distância* (INED), a public and regulatory body that among other assignments must define policies, undertake the audit and review of DL institutions and programs at all levels of education (Conselho de Ministros 2006). Another body is called *Conselho Nacional de Avaliação de Qualidade* (CNAQ) which is responsible for academic compliance of programs. The third is *Direcção Nacional do Ensino Superior* (DINES) which caters for the legal aspects of the establishment and running of all higher education academic institutions. Figure 1 shows the relationship between the three key stakeholders in the full cycle of DL accreditation.

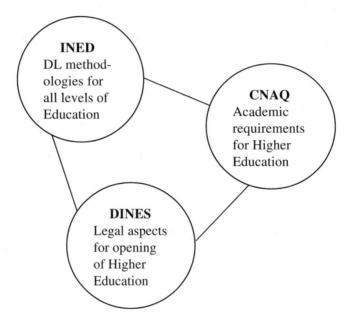

Fig. 1. The three key bodies for full distance learning accreditation process in Mozambique.

In 2009, the Regulation of ODL was approved and it is applicable to all levels and types of education. According to the regulation, accreditation is defined as quality certification awarded by the accrediting body, INED, based on the results of the external evaluation to the institution, course or study program (Conselho de Ministros 2009). The accreditation system in Mozambique consists of seven key DL dimensions applied in a systemic approach to evaluate both institutions and programs. Each dimension is given a relative weight resulting from the need to respond to the type of irregularities most frequently committed by institutions. For example, the dimensions' *pedagogical organization*, *study materials* and *student support services* hold a greater percentage (20%) each, than the *institutional strategy for distance education* (6%) and *monitoring and evaluation* dimension (7%); as they are the areas that have the greatest impact in the provision of DL courses nowadays in Mozambique. Below the seven key dimensions for quality assessment in DL and their respective relative weights are presented (INED 2014):

Dimension 1:   Institutional strategy for distance education (6%)
Dimension 2:   Pedagogical organization (20%)
Dimension 3:   Physical and technological resources (10%)
Dimension 4:   Study materials (20%)
Dimension 5:   Student support services (20%)
Dimension 6:   Human resources (17%)
Dimension 7:   Monitoring and evaluation (7%)

3   Distance Learning and Smart Territory Development

What is known today as DL goes back as far as the 1880s with the emerging of correspondence education. People willing to study at home or at work were able to do so through printed material, which reached them by postal services that took advantage of the railways network (Moore and Kearsley 2007). As other media became available and used to offer courses at distance, correspondence education became known as distance education (Holmberg 2003). "Distance Education is a suitable term to bring together both the teaching and learning elements of this field of education" (p. 38) (Keegan 1996). There is therefore an element of teaching and learning in this mode of education, happening separately from each other being this the main characteristic of this mode of education. In DL "(…) teachers and students are in different places for all or most of the time that they teach and learn. Because they are in different places, in order to interact with each other they are dependent on some form of communications technology" (Moore and Kearsley 2012, 1).

World over, DL has been used for different reasons. For instance, DL has been an opportunity for those with jobs, family responsibility and social commitments, but they can study at home or at work at their own time. People living in remote areas or where there is no school at all can as well benefit from DL and, in that way, not prevented from education. In some countries, the educational problems have part of their solution embracing DL with many countries in Africa establishing structures to run educational programs through this mode of education (Perraton 2012).

Mozambique introduced the National Education System in 1983 and adjusted it in 1992 under the law 6/92 to meet the new political and social transformations that the country was faced with. The National Education System contemplates DL as part of the System and can be used on its own or as component of a certain educational program.

Before independence, Mozambique experienced DL programs run from Portugal. The programs were very practical and people learned skills through short courses such as cooking, sewing, electricity, radio and TV repair among others. People acquired knowledge and skills that enabled them to deal with practical issues of daily life providing solutions for them and people surrounding them contributing in that manner to the betterment of their lives and families and ultimately of the community in which they were living. Through DL and without having to leave their environment, families and community, people learned a profession that enabled them solving practical problems and also generate some income for their families.

The first DL experience under the National Education System was with the in service training of unqualified primary school teachers. The program started in 1984 through printed material, radio and occasional face-to-face sessions. Around 1,200 teachers from six (6) provinces were registered for training with the program. Due to the flexibility of DL these teachers did not need to leave their schools to attend training. While working they took training that enabled them to put into practice in the classroom the new skills, which contributed to improve their professional capacity and facilitating students learning.

Since then, many other programs followed that experience. In 2013 there were around 14 institutions offering DL programs and courses, 52.389 students and around 50 courses (Ministério da Educação 2013). DL is enabling people to get education without having to abandon their jobs, families and community; and in some cases not having to miss education because of the lack of educational opportunities locally.

Through DL people stay in their villages, towns and communities avoiding travelling and moving to other places to get education. This helps families remaining together and reduces the burden of the costs of education in other places far from their region providing them the opportunity to contribute to the development of where they live. While in Mozambique in every town and village there is a primary school, the same does not happen with secondary schools or higher education institutions, which are mainly located in provincial capitals and in some major towns.

Education is key to development. In "(…) the late 1950s and early 1960s there was general agreement among politicians, educational and social planners, and scholars that education was a key agent for moving societies along the development continuum" (Fagerlin and Saha 1989, 40). In the beginning of the 1960s the world started investing in education (formal and informal) with the believe that by educating and training people would impart skills, knowledge, attitudes and motivation to contribute to economic and social development (Psacharopoulos and Woodhall 1985).

Since the early days of independence, Mozambique invested tremendously in education and, as a result, the education system grew considerable. From 1974/1975 to 2016, primary schools went from 5,261 to 12,527; secondary schools went from 9 to 703; vocational and professional schools went from 29 to 120 and at the tertiary level from one (1) institution to 49 institutions. In the same period, the number of students in total went from 709,299 to 7,276,981 (Ministério da Educação e Desenvolvimento Humano 2016).

Although the tremendous improvement in education since independence more needs to be done, so that access to education could become available to most of the population. The expansion of education, obviously, requires resources. Resources are not always available which could pose a problem in the provision of education for the population.

The Education Strategic Plan (ESP) 2012–2016, which will be prolonged until 2019 to ensure the fulfilment of the planned activities, highlights "(…) the continued expansion of the Education System, within the possible limits to guarantee the quality of the educational services offered, through exploring several delivery modalities, including distance learning and benefitting from the potential of new technologies" (Ministry of Education 2012, 2).

Education is crucial for development. With education people are better prepared to contribute with their knowledge and skills for the development of the society and the country. The National Education System in Mozambique recognizes DL as a delivery mode of education opening opportunities for everyone to learn no matter the condition ensuring, therefore, expansion and access to education. Through DL Mozambique is delivering primary education, secondary education, vocational education and higher education programs to those in need. With these programs, people are getting the knowledge and skills without having to abandon their jobs, leaving families and abandon their communities to get education. They are acquiring knowledge and skills to deal with the challenges they are faced with and come up with well-informed solutions contributing to their economic and social development as well as of the country. DL with its main characteristic of being a flexible mode of learning can contribute to the local smart development by enabling people to remain in their environment, villages and towns avoiding pressure of the already overcrowded urban areas in search for academic institutions.

4   Challenges to Quality Public Policies Towards Territory Development

In general, the challenges of DL public policies leading to the development of the territory accompany the stage of socio-economic development of a space. Although diversification of sources of knowledge and the way it is accessed is a valuable form of preparing people to actively participate in the development of a territory; it is still a challenge to have acceptable mechanisms that can guarantee quality of education provided. The extension of sources and forms of knowledge are somehow a major challenge worldwide for the development of public policies conducive to the provision of quality DL.

In the context of developing countries and Mozambique, in particular, although political enabling environment is visible there are still major challenges in the provision of the basic technological infrastructure, such as power supply, bandwidth and equipment; qualified personnel to deal with technology; experts on the DL quality public policies and financial resources for a number of initiatives including research.

It is also a challenge to develop policies that promote development of territories since it would be necessary to avoid the "one size fits all" approach but taking into account particular contexts to allow people not to leave their places of residence and/or work in search of educational opportunities, which are often found in the big cities, as in the case of Mozambique. The rural exodus very often contributes not only to the delay in the development of the countryside, but also to the demographic pressure in urban areas. The challenge is to make a public policy caters for different realities in order to achieve the desired development of a space made up of differentiated contexts.

The spectrum above briefly presented usually is mitigated through definition of strategies in order to focus and overcome the main challenges. In Mozambique, the issue of quality is one of the three pillars of education. The ESP recognizes that "the sector will continue to improve the quality and relevance of post-primary education to strengthen its role in the economic, social and political development of our society" (Ministry of Education 2012, 33). In addition, as recognized in the DL Strategy for the period 2014–2018 (Ministério da Educação 2013) the government advocates for a DL system that guarantees citizens access to the different levels and types of education through expansion of institutions and programs that meet the needs of socio-economic development with quality and equity. With this approach it is expected from the government policies for quality assurance in DL to cater not only for people well-being but also for smart territory development nationwide as well as beyond borders

Due to the rapid increase of DL providers, including the diversification of courses offered, as a response to the government strategy, there was a need to regulate the use of this educational modality in the country. To this end, the government created INED, in 2006, as a public and regulatory body. Among other assignments must create and develop the system of quality assurance and accreditation in ODL (Conselho de Ministros 2006). Regarding government quality assurance mechanisms for DL to face accreditation of transnational DL programs, it is suggested that countries work closely in the recognition of a program in the country of origin.

In the near future, it is worth to mention that studies need to be carried out to recommend what DL public policy should address to accommodate different needs of different people in different contexts, as in the case of Mozambique. With quality public policies that recognize context differentiation, DL would contribute to the promotion of territory development since people would access quality education and training remaining in their communities, and thus responding to the local socio-economic demands.

Acknowledgments. This article reports research developed within the PhD Multimedia in Education (University of Aveiro, Portugal), integrated in the PhD Program Technology Enhanced Learning and Societal Challenges, funded by Fundação para a Ciência e Tecnologia, FCT I.P. – Portugal, under contracts # PD/00173/2014, # PD/BI/113830/2015 and # PD/BI/113826/2015.

Acknowledgments are extensive to the Ministry of Education and Human Development (MINEDH) and the Ministry of Science and Technology, Higher and Vocational Education (MCTEST) in Mozambique for the support in providing valuable data for this study.

References

Bates T (2008) Transforming distance education through new technologies. In: Evans T, Haughey M, Murphy D (eds) International handbook of distance education, First edit. Emerald Group Publishing Limited, UK, pp 217–235

Conselho de Ministros (2006) Decreto n. º 49/2006 de 26 de Dezembro Estatuto Orgânico do Instituto Nacional de Educação à Distância (INED). Publicado no Boletim da República, Maputo, Moçambique

Conselho de Ministros (2009) Decreto nr 35/2009 de 7 de Julho. Regulamento do Ensino à Distância. Publicado no Boletim da República, Maputo, Moçambique

Fagerlin I, Saha LJ (1989) Education & national development: a comparative perspective, 2nd edn. Pergamon Press, London

Green D (1994) What is quality in higher education? Society for Research into Higher Education (SRHE) and Open University Press, London. https://eric.ed.gov/?id=ED415723

Harvey L (2006) Understanding quality. In: Purser L (ed) Introducing Bologna objectives and tools: UA Bologna handbook: making Bologna work

Harvey L, Green D (1993) Defining quality. Assess Eval High Educ 18(1):9–34

Holmberg B (2003) Distance education in essence: an overview of theory and practice in the earlier twenty-first century. Edited by University of Oldenburg, 2nd edn. BIS, Oldenburg

INED (2014) Acreditação de Instituições e de Cursos de Educação à Distância (EAD) em Moçambique. - Manual de Procedimentos do Provedor -. Edited by Instituto Nacional de Educação à Distância (INED). Vesrão 1. Ministério da Educação, Maputo

Keegan D (1996) Foundations of distance education, 3rd edn. Routledge, London

Ministério da Educação (2013) Estratégia da Educação à Distância 2014–2018. Conselho de Ministros, Maputo, Moçambique

Ministério da Educação e Desenvolvimento Humano (2016) Estatísticas de Educação. Maputo, Moçambique

Ministry of Education (2012) Education strategic plan 2012–2016. Conselho de Ministros, Maputo, Moçambique

Moore M, Kearsley G (2007) Educação a Distância: Uma Visão Integrada. São Paulo

Moore M, Kearsley G (2012) Distance education: a systems view of online learning

Ossiannilsson E, Williams K, Camilleri AF, Brown M, International Council for Open and Distance Education (ICDE) (2015) Quality models in online and open education around the globe: state of the art and recommendations. Online Submission

Perraton, H (2012) Theory, evidence and practice in open and distance education. Commonwealth of Learning and Athabasca University

Peters O (2004) Digitised learning environments: new possibilities and opportunities. In: Peters O, Distance education in transition. New trends and challenges, pp 57–70

Psacharopoulos G, Woodhall M (1985) Education for development: an analysis of investment choices. Edited by World Bank. Oxford University Press, Whashington

UNDP (2015) Sustainable development goals 2015–2030. J Chem Inf Modeling. UNDP. doi: 10.1017/CBO9781107415324.004

Vlăsceanu L, Grünberg L, Pârlea D (2007) Quality assurance and accreditation: a glossary of basic terms and definitions. Edited by Melanie Seto and Peter J Wells. UNESCO

Smart Learning Resources

Participatory Design of Tangibles for Graphs: A Small-Scale Field Study with Children

Andrea Bonani[✉], Vincenzo Del Fatto, Gabriella Dodero,
Rosella Gennari[✉], and Guerriero Raimato

Faculty of Computer Science, Free University of Bozen-Bolzano,
Piazza Domenicani 3, 39100 Bolzano, Italy
`{abonani,vincenzo.delfatto,gabriella.dodero,`
`guerriero.raimato}@unibz.it, gennari@inf.unibz.it`

Abstract. Algorithmic thinking is at the heart of computational thinking: it requires to abstract a problem and to model it, as well as to specify a sequence of instructions for solving it, that is, an algorithm. In many countries, computer science education in primary or secondary schools is moving towards computational thinking and, partly, algorithmic thinking education. This paper supports the idea that algorithmic thinking should be taught from primary schools, through physical activities, without sitting in front of a computer screen, and by exploiting Internet of Things technologies, specifically, by using inter-connected interactive tangible objects. The paper focuses on the design of such tangibles for teaching graph algorithmic thinking. It shows in what sense their design was participatory and followed an action-based re- search approach, moving from the context of use analysis to iterative design sessions and field studies with tangible prototypes, used by teachers or children. The paper presents one of such tangibles and field studies with children. The paper ends by reflecting on the design process of tangibles for graph algorithmic thinking.

Keywords: Algorithmic thinking · Tangible · Interaction design

1 Introduction

Algorithmic thinking (AT), which is at the hearth of the best known *computational thinking (CT)* (e.g., Maloney et al. 2010), requires to abstract the essentials of a problem and to model it, so as to give step-by-step instructions (an algorithm) for its resolution (Futschek 2006). In many countries, CT and its core, AT, are increasingly being considered for training programmes for teachers and education curricula of schools, from primary schools onwards. The use of interactive tangibles for teaching AT potentially fosters the interplay between abstraction and concreteness, by stimulating different learning modalities, following a constructive approach to learning (Futschek and Moschitz 2011). This paper focuses on interactive tangible objects for graph algorithmic thinking (briefly, tangibles for graph AT), for primary and middle school contexts.

Simply put, a(n undirected) graph is an ordered pair $G = (V,E)$, with $V = \{v_1,...,v_n\}$ a set of nodes, and $E = \{v_i,v_j | v_i,v_j \in V\}$ a set of edges between pairs of nodes. Graph AT

Ó. Mealha et al. (eds.), *Citizen, Territory and Technologies: Smart Learning Contexts and Practices*,
Smart Innovation, Systems and Technologies 80, DOI 10.1007/978-3-319-61322-2_16

requires to model a problem with a graph, to understand its properties and to envisage algorithms for deciding on them. Example problems, familiar to children, are given by social networks, such as Facebook, with profiles of users, related through friendship relations. Profiles can be modelled as nodes of a graph, and edges between nodes represent friendship relations. Problems are: is there a way to connect a user profile, A, to a user profile, B, through mutual friends? That amounts to the verification of a property: whether there is a path in the Facebook graph between A and B. Given our set of profiles of users, are they all connected through mutual friends? That means verifying if the graph is connected, another graph property. If not, how can one connect all the given profiles of users? In other words, how can one devise an algorithm for that? Understanding graph properties and modelling are prerequisites for understanding and devising graph algorithms.

This paper builds on the idea that tangibles for graph AT can enhance the learning of graph AT, conceived as a continuous contextualised process as in constructivism (e.g., Fosnot 2005). The design of such tangibles is thereby highly complex: it should release evolving tangibles and explore their users' appropriation in field studies, according to their specific learning contexts. Participatory design or, more generally, action- based research approaches, which recommend the deployment of exploratory tangibles in real-world contexts and assess their usage over time, could help design tangibles that are informed by repeated usage practices. This paper promotes the idea of using participatory design for tangibles for graph AT.

The paper starts by providing background information in Sect. 2: related work, the reference learning theory and participatory design approach. It moves on presenting the participatory process of tangibles for graph AT, and the most recent prototype in Sect. 3. Section 4 presents results from the latest small-scale field study with a teacher, primary and middle school children. Section 5 concludes the paper and recaps lessons learnt for future work.

2   Background

2.1   Related Work in AT or Graph Education for Schools

Perhaps the best known approaches to AT education are based on coding, and programming environments for children, e.g., the aforementioned Scratch. Proposals, such as CS-unplugged by Bell et al. (2008), instead, teach CT without computers and with pencil-and-paper interventions, which require physical activities. CS-unplugged has inspired similar research work, e.g., algomotricity (Bellettini et al. 2014), with puzzles (Lamagna 2015). In particular, teaching of graph modelling and algorithms through physical non-interactive objects has also been the subject of research in the recent past, in school contexts, e.g., (Gibson 2012). Such pencil-and-paper interventions are predominant in countries or schools without sufficiently many computer labs, which is the case in the geographical area we work in.

Relevant related work for AT education is then in the area of interaction design of tangibles (briefly, ID), e.g., (Bers et al. 2014; Lee et al. 2011). Teaching AT through tangibles has received increasing attention in recent years. AT has been introduced, for instance, with a haptic model (Capovilla et al. 2013). AT activities with tangibles and involving multiple learners are fewer. The main reference is (Gennari et al. 2016), which describes a gamified tangible for primary schools, BALA, for the scaffolding of the bubble sort algorithm.

2.2 Main Reference Learning Theory

This paper moves along the lines of fostering learning of graph AT through physical activities, together, according to the learning contexts. Its main reference learning theory is constructivism in a sociocultural approach (e.g., Fosnot 2005). According to constructivism, learning is not absorbing facts, but an active dynamic knowledge-construction process. In a constructive learning experience, a learner is usually given tools to construct knowledge, possibly through different senses.

Constructivists also purport that learning is tied to learners' experience and con- text; learning is based on learners' existing knowledge, and needs to be connected to their context so that people perceive learning as meaningful for them.

While classic constructivism emphasises individual knowledge construction, sociocultural constructivism stresses the importance of interactions with others for co-constructing knowledge and mutual learning: teachers, experts, or peers. Learners can tackle difficult learning tasks with others, as they move through their zone of proximal development, in which scaffolding gradually changes and fades away.

In brief, in sociocultural constructivism, it is through different modalities and interactions with diverse tools and people that learners can truly learn.

2.3 Main Reference Design Approach

The design of tangibles for graph AT has several dimensions to consider, e.g., whether: tangibles can support the scaffolding of graph AT, in a constructive manner; tangibles can be used by their users (usability) and inspire novel usages.

That motivates the adopted design approach, which is participatory and based on action-research. This approach tries to involve users in a continuous spiralling process: designers release evolving prototype solutions for the learning context; actions in the field are organised (that is, field studies), in which users use prototype solutions and unveil design and learning possibilities; field studies are reflected over and novel solutions are advanced, e.g., (Gennari et al. 2017a, b). All participants in the process should benefit from it; users and designers can continuously contribute to the development of tangibles, and all they can mutually learn through the process, in line with sociocultural constructivism. How the participatory approach was applied for tangibles for graph AT is explained next.

3   The Design Process of Tangibles for Graph AT

Participatory design with action research, introduced above, is adopted in this research for developing evolving tangibles for graph AT. The design process releases exploratory prototype solutions, which are used in studies with designers and users; these are teachers and 9–13 years old pupils from primary and middle schools.

Specifically, the design process started with a context of use analysis, which triggered the first design ideas. Designers released alternative design mock-ups and then prototypes of tangibles for graph AT. They also delivered usage scenarios for primary and middle schools. The figure depicts an usage of such interactive tangibles. It shows a group of four learners who are building a graph with a set of tangibles for nodes and edges, besides a tangible "confirmation button" that children can use to signal the conclusion of their task. A class teacher monitors students' activities through her or his own interface, on a computer.

The figure also shows some tangibles for graph AT of the most recent prototype: three tangibles for nodes (wood boxes with LEDs); tangible for edges (Ethernet cables); a confirmation button (white and circular, with a blue button on top).

The prototype implements a distributed client-server architecture. Server and clients communicate through a WiFi connection. The server is a computer that verifies graph properties, and implements the graphical interface for teachers. Clients, which are the tangibles for children, interact with the server and children through micro-electronic components.

From the interaction viewpoint, the design of tangibles is modular and vertical: it implements few but relevant interaction functionalities to explore and rapidly adapt according to children's usage and ideas. Edges, which are Ethernet cables, are just passive links, and presently provide no interaction. The confirmation button can only be pressed by users for signalling that they concluded their work and are awaiting the feedback of nodes. The main interactive tangibles for users are nodes.

Each node is equipped with an RGB node LED and three RGB edge LEDs, besides other microelectronics components. RGB node LEDs are activated for giving specific types of feedback, e.g., RGB node LEDs of a strongly connected component switch on with the same colour. Each node has three sockets for cables, and in parallel three RGB edge LEDs. The edge LEDs are activated when a cable is inserted in its socket, or for delivering other types of feedback concerning edges, e.g., in a simple graph there is only one edge between a pair of nodes (Fig. 1).

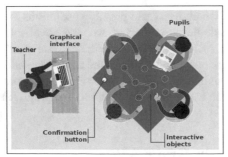

Fig. 1. Envisioned usage (left) and some of the prototype tangibles for AT (right).

4   Field Study

Prototype tangibles for graph AT, made of cartoon, were used by experts of interaction design and then teachers. The usage sessions with teachers served to uncover potential usability issues and unexpected usage scenarios. In particular, teachers found the interaction design choices clear for their pupils. A teacher was concerned with the fragility of the nodes, cartoon made, and hence they were made with wood in the latest prototype for children. Usage sessions with teachers suggested that tangibles for graph AT could be used to teach pupils properties of binary relations on a set (the graph edges), which are part of traditional curricula—reflexivity and asymmetry. This scenario was implemented in the current prototype, but not tested yet.

The field study with the latest prototype of tangibles for graph AT aimed at inspecting whether children would grab graph AT ideas with the prototype, at dis- covering further potential usability issues as well as at unveiling novel design ideas with children. The study took in total 20 min and was run in an informal learning context. Its main results are described next.

4.1   Study Design

Participants and Roles. A designer and usability expert was present with a teacher, who acted as moderator and expert of learning contexts, as well as of users of the developed technology. Participant children were 3, one from primary schools, 9 years old, and the others from middle school, 13 years old.

Goal and Material. The study goal was gathering qualitative data useful for the design in relation to graph AT understanding through tangibles, usability issues and children's design ideas for the prototype tangibles, like in (Di Mascio et al. 2014; Gennari et al. 2017a, b). The study was thus formative and with the most recent prototype of the tangibles for graph AT; see the figure. The study also used scenarios, which are briefly described as follows.

Scenarios. Scenarios were incremental for the scaffolding of graph AT, in line with the chosen learning approach (see Sect. 2) and related work in AT education, e.g., (Gibson 2012). All scenarios used the Facebook social network metaphor; the teacher progressively unveiled scenarios and asked probing questions concerning graph AT. The first scenario was the basic one. It introduced children to the prototype nodes and edges; the teacher asked children how the prototype could be used to model the social network. In the second scenario, the teacher guided children to construct a simple graph with two strongly connected components. Then the teacher asked children to push the confirmation button, and the node LEDs of the connected components coloured differently (see the figure). The teacher posed questions concerning children's understanding of the graph properties. In the third scenario, the teacher asked children how they would construct a connected graph (in which all profiles are reachable through friends), in order to understand what algorithm they would devise.

4.2 Study Results

Graph AT. Children were incrementally introduced to the prototype and all three scenarios. As for the first scenario, all children understood how to model the social network with the prototype nodes and edges: nodes were used to model user profiles of the social network, and edges for friendship relations. As for the second scenario, all children managed to physically construct the graph under teacher's instructions, distributing work (who connects what). When the confirmation button was pushed, they commented that nodes with different colours "were for people who did not know each other". When the teacher moved them to the third scenario, all of them opted for the minimal number of nodes to connect in order to construct a connected graph (any two nodes of the two components).

Usability Issues and Uncovered Design Ideas. Children found the interaction design choices usually clear for them: the green for LEDs, meaning that the choice was correct; edge LEDs blinking with different colours, one per strongly connected component, if the graph is not connected. However, all children found the blink timing too short: they wished to have longer time to watch LEDs and reason on strongly connected components. Children suggested to have colours different than green for strongly connected components—green should be reserved for "ok". They also wanted designers to introduce sound effects: a "happy sound" when the desired property is satisfied (e.g., connectedness) and a "fun" sound otherwise.

From the product design viewpoint, several issues emerged. For instance, boxes should be round to allow users to easily turn them and see all edge LEDs. Cables were not always easy to manage for the younger child, and overlapping edges were confusing children.

5 Conclusions

This paper introduced the idea of tangibles for graph AT, for scaffolding the learning of graph AT in an active multimodal experience, as recommended by constructivism. It motivated and presented a participatory design process for them, with an action-research approach. This spirals through exploratory prototypes of tangibles for graph AT and field studies; then users (teachers and children) and designers use tangibles and co-discover design and learning possibilities. The paper presented the most recent prototype of tangibles for graph AT, and the related small-scale formative field study with children. Tangibles were used to convey the modelling of social network problems with graphs, and the scaffolding of a simple graph algorithm.

According to the study findings, children could grasp basic properties of graph, model problems with the tangibles and devise a strategy for an algorithm that builds a connected graph. Usability issues and few design ideas also emerged in the study. The results of the study are limited in scope and yet interesting for the participatory design of tangibles for graph AT, as discussed in the remainder.

Design results were limited to specific features of tangibles. Possible reasons for that are the short time of the study (20 min), and the fact that children could express their design ideas mainly verbally. Future studies should allow children to freely play with the tangibles, as much as they wish, and give them material to express their design ideas differently than in words or through designers' observations. Another explanation is that, being vertical prototypes, tangibles were (on purpose) limited in interaction design functionalities; however, that allowed researchers to rapidly and easily assess their usage by children. The results of the study are limited in terms of participants (a teacher and three children from different schools) and investigated scenarios. Anyhow, the results are useful in that they indicate how such tangibles can be used in primary and middle schools to enhance graph AT.

References

Bell T, Alexander J, Freeman I, Grimley M (2008) Computer science without computers: new outreach methods from old tricks. In: Proceedings of the 21st Annual Conference of the National Advisory Committee on Computing Qualifications

Bellettini C, Lonati V, Malchiodi D, Monga M, Morpurgo A, Torelli M, Zecca L (2014) Extracurricular activities for improving the perception of informatics in secondary schools. In: Informatics in schools. Teaching and learning perspectives. Springer, pp 161–172

Bers MU, Flannery L, Kazakoff ER, Sullivan A (2014) Computationalthinkingandtinkering: exploration of an early childhood robotics curriculum. Comput Educ 72:145–157

Botero A, Kommonen K, Marttila S (2011) Design from the everyday: continuously evolving, embedded exploratory prototypes. In: Proceedings of DIS 2010. ACM, pp 282–291

Capovilla D, Krugel J, Hubwieser P (2013) Teaching algorithmic thinking using haptic models for visually impaired students. In: LaTiCE 2013. IEEE. pp 167–171

Di Mascio T, Gennari R, Melonio A, Tarantino L (2014) Engaging new users into design activities: the TERENCE experience with children. Springer International Publishing Cham, pp 241–250

Fosnot CT (2005) Constructivism: theory, perspectives, and practice. Teachers College Press, New York

Futschek G (2006) Algorithmic thinking: the key for understanding computer science. In: Informatics education–the bridge between using and understanding computers. Springer, pp 159–168

Futschek G, Moschitz J (2011) Learning algorithmic thinking with tangible objects eases transition to computer programming. In: Informatics in schools, contributing to 21st century education. Springer, pp 155–164

Gennari R, Del Fatto V, Gashi E, Sanin J, Ventura A (2016) Gamified technology probes for scaffolding computational thinking. In: De Angeli A, Bannon L, Marti P, Bordin S (eds) COOP 2016: Proceedings of the 12th international conference on the design of cooperative systems, 23–27 May 2016, Trento, Italy. Springer, Cham, pp 303–307

Gennari R, Melonio A, Raccanello D, Brondino M, Dodero G, Pasini M, Torello S (2017a) Children's emotions and quality of products in participatory game design. Int J Hum Comput Stud 101:45–61

Gennari R, Melonio A, Rizvi M (2017b) The participatory design process of tangibles for children's socio-emotional learning. In: Proceedings of IS-EUD 2017. LNCS, vol 10303. Springer

Gibson JP (2012) Teaching graph algorithms to children of all ages. In: Proceedings of ACM ITiCSE 2012, p 34

Halskov K, Hansen NB (2015) The diversity of participatory design research practice at PDC 2002–2012. Int J Hum Comput Stud 74:81–92

Lamagna EA (2015) Algorithmic thinking unplugged. J Comput Sci Coll 30(6):45–52
Lee I, Martin F, Denner J, Coulter B, Allan W, Erickson J, Malyn-Smith J, Werner L (2011) Computational thinking for youth in practice. ACM Inroads 2(1):32–37
Maloney J, Resnick M, Rusk N, Silverman B, Eastmond E (2010) The scratch programming language and environment. ACM Trans Comput Educ (TOCE) 10(4):16
Wing JM (2006) Computational thinking. Commun ACM 49(3):33–35

The Diorama Project: Development of a Tangible Medium to Foster STEAM Education Using Storytelling and Electronics

Sanne Cools[1], Peter Conradie[1,2(✉)], Maria-Cristina Ciocci[3], and Jelle Saldien[1]

[1] Department of Industrial Systems Engineering and Product Design, Ghent University,
Graaf Karel de Goedelaan 5, 8500 Kortrijk, Belgium
{Sanne.Cools,Peter.Conradie,Jelle.Saldien}@UGent.be
[2] imec-mict-UGent, Miriam Makebaplein 1, 9000 Ghent, Belgium
[3] Department of Pure Mathematics, Ghent University, Krijgslaan 281, 9000 Ghent, Belgium
Cristina.Ciocci@UGent.be

Abstract. Children of the 21st century grow up in a world full of information and technology. Education should equip them with useful skills and competencies, allowing them to actively and effectively take part in a globalised society. Teachers feel the need for educational tools that support innovative teaching. To this end, this paper describes the development of The Diorama Project. This series of trans-disciplinary workshops combines familiar subjects, like language and art, with new topics such as programming and electronics, to foster valuable skills and knowledge in a more fun and tangible way. Pupils team up to write, record and tinker a story. Programmable electronics let their theatre plays come alive. An open source platform provides all the information for teachers to organise the workshops by themselves. They can use it to share their experience and knowledge with colleagues worldwide.

Keywords: STEAM education · 21st century skills · Play

1 Introduction

Recent years has seen a renewed focus on skills such as critical thinking and reasoning as part of education, so called 21st Century skills. While they are perhaps not new (Willingham 2010), they can be viewed as *newly important* (Silva 2009). Yet, despite renewed focus, according to the World Economic Forum (2015) all too often pupils are not attaining the essential skills they need to prosper in this century. Providing teachers with ready-to-use educational tools, is necessary to tackle this problem (Mataric et al. 2007; Thijs et al. 2014).

While the industrial society of the last century aimed at mass consumption focusing on motorization (Masuda 1980), modern society can be viewed as a 'knowledge society' (Anderson 2008; Dede 2010; Halász and Michel 2011) which refers to economic systems where ideas or knowledge function as commodities (Anderson 2008). Beside dynamic changes in the types of jobs demanded in the knowledge society (Reich 1991), we also

© Springer International Publishing AG 2018
Ó. Mealha et al. (eds.), *Citizen, Territory and Technologies: Smart Learning Contexts and Practices*,
Smart Innovation, Systems and Technologies 80, DOI 10.1007/978-3-319-61322-2_17

face the challenge to educate young people for jobs that don't exist yet (Dede 2010; Voogt and Roblin 2012).

The World Economic Forum (2015) proposed 16 skills, grouped into three broad categories (Foundational Literacies, Competencies and Character Qualities, see Fig. 1). 21st Century Skills can be added as new subjects or new content within traditional subjects. However, most of them are not directly linked to a specific field but relevant across many fields.

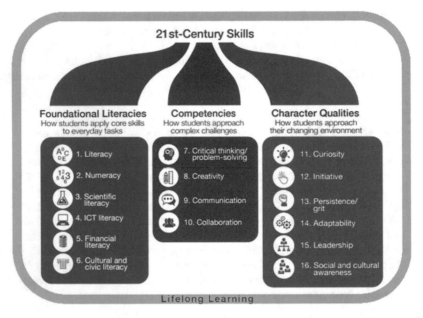

Fig. 1. 21st Century skills, according to the world economic forum (2015)

1.1 Learning Through Playing

To engage pupils, especially children, the inclusion of play in the learning process can be helpful. By combining what children learn and what they do for fun, exciting learning experiences can occur (Johnson and Thomas 2010). Children develop creative thinking techniques that are important to many fields, but particularly useful for learning science and engineering (Mcgrath and Brown 2005). Ferrari, Cachia and Puni (2009) have identified that creativity and innovation have strong links with knowledge and learning. Creativity is seen as an essential skill that leads to knowledge creation and the construction of personal meaning

Tangible mediums as learning tools also benefit the learning process. Resnick (2006) found that familiar objects used in unfamiliar ways can spark children's interest in learning new subject matter. They let children use food as musical instruments, for example, to explore the resistance of different substances. Children were more comfortable experimenting and exploring while playing with familiar materials. At the same time, they were more triggered when unexpected things happened with these objects.

1.2   The Role of Technology

Most 21st century competencies can either be supported or enhanced by information and communications technologies (ICT) (Ananiadou and Claro 2009). The use of technology can also expand children's understanding of traditional school subjects, such as literacy, science and mathematics. Besides enhancing learning experiences, technologies can foster innovative teaching (Roschelle et al. 2000). It gives teachers the opportunity to reconsider how courses and learning activities are organized (Blumenfeld et al. 1991; Beetham and Sharpe 2013).

Multimedia, such as video and animation can be used during class to explain complicated concepts. The purpose of technology is to facilitate learning, but ICT and other technological mediums can only be relevant when they are used as learning tools, and not just for the sake of using technology (Niess 2005). Further, technology will only enhance the learning experience when it suits the age and the proficiency level of the pupils.

The implementation of new competencies and new education strategies, but also the use of technology in the classroom, can be major challenges for teachers (Niess 2005). To engage in this educational change, teachers need to feel competent and value what they are teaching (Blumenfeld et al. 1991; Ananiadou and Claro 2009). They should spend sufficient time on supporting their pupils and are also expected to have mastered all the skills and knowledge themselves (Gordon et al. 2009). To bring these theories into practice, teachers should be supported with educational tools (Mataric et al. 2007; Thijs et al. 2014).

An emerging approach for teaching 21st Century Skills is STEAM (Ge et al. 2015), explained by Yakman as (2008) "Science and Technology interpreted through Engineering & the Arts, all understood with elements of Mathematics". Art broadens pupils' horizon, stimulates them to imagine the impossible and follow their intuition, and is crucial to support STEM skills (Tarnoff 2010). As noted by Peppler (2013), STEAM-powered education can additionally play a role to correct the gender balance currently present in engineering disciplines (Barnard et al. 2012).

Most children decide whether to choose a STEM related career before the age of 14, based on early experiences (Tytler et al. 2008). Functional and educational toolsets developed for pupils and teachers, can be of great support for educators to foster early interest in STEM related disciplines.

A child's learning process especially benefits from interaction with tangible objects. Lego Education WeDo 2.0 (Lego Education 2017) and littleBits (2017) are two popular education tools to bring STEAM subjects – especially electronics and coding – in primary schools.

2   Introducing: The Diorama Project

To address the challenges of introducing 21st Century Skills in the classroom, with an emphasis on art and creativity, we introduce the Diorama Project, a series of transdisciplinary workshops where familiar subjects, like languages and art, are combined with new topics such as programming and electronics. The overall goal of the workshop is to make a table top mini-theatre (see Fig. 2), using a construction kit with four empty stages. Three to five pupils collaborate to create a mini-theatre play by building the mini-theatre (engineering), writing a script, recording it (spoken word, ICT) and tinkering with the stages, props and characters (visual art, engineering). Finally, with Microduino mCookie's programmable modules (Microduino Studio) they can let their theatre plays come alive (technology, math).

Fig. 2. Example of a mini-theatre

The project thus integrates many of the 21st century skills described by WEF (2015). Critical thinking, creativity, communication and collaboration are strongly encouraged. Besides, a long-term project also requires pupils to show persistence, leadership and take initiative to reach their goal. Teachers choose when their Diorama Project starts and ends, spreading it over an entire month or one week. An open source platform provides them with all the information to organize the workshops by themselves, or they can ask help of people familiar with computer science and programming.

For this project Microduino mCookie smart modules are used to control the mini-theatre. Children can program these modules with Blocky4Arduino. This is a web-based, visual programming application for Arduino developed by Ingegno. The visual interface is colourful, easy to use and accessible for children of all programming levels.

In the following section, we discuss the overall development process of the Diorama Project, followed by our main lessons learned.

3   Developing the Diorama Project

The experience of children with early versions of the Diorama kit took a central place in its development. The product can only become functional and satisfying when users are involved throughout the whole process of designing. Therefore, user centred workshops were organized frequently during the entire design of the tangible medium and the content development of the project.

In total, four iterative and interactive workshops were conducted during the design process. The goal of the workshops was to identify any usability problems, and determine the participant's satisfaction while building their mini-theatre.

Nearly 40 children were involved during the development of the project. In total, participants developed 10 different mini-theatres. After workshop 2 and 4 the participants were asked to fill in a questionnaire. Questions centred around what pupils found challenging, what they learned or their experience with various aspects of designing the theatre, such as programming or scripting the dialogue.

Every workshop was observed to make continuous improvements on the mini-theatre prototypes, programming platform, project plan and activities. Below we provide a brief overview of each workshop.

Workshop 1: Ten children from 10 to 12 years old participated. In teams of 3 to 4 members, the children gathered information about their topic, tinkered the stages of the mini-theatre and recorded the text. They were introduced to the Arduino IDE. Finally, they gave a presentation with their mini-theatre for their parents and siblings.

Workshop 2: A mini-theatre project was organised with the 5th grade (age 10–11) of a local primary school. Ten girls and eleven boys took part in a workshop of three hours. They made teams of four to five members. The teacher asked each group to think of a fictive story for their mini-theatre. She led the story writing activity. Next, the children did sound recording and tinkering under the supervision of the designer. Programming would be done in a later workshop.

Workshop 3: A small class of 6th graders (age 11–13) was invited for a mini-theatre workshop at the local library. The workshop covered an entire school day (5 h). Four girls and three boys took part. The children were asked to base their story on a book that they could choose from the library. Since the time was limited they were suggested to choose a picture book. They went through all activities of the project, including programming with a new visual programming platform.

Workshop 4: A final workshop was organised with the same group of children that took part in user workshop 2. The main focus of this workshop was programming with the new visual programming platform. Each group could use one laptop. The workshop lasted for three hours again. The teams also continued tinkering their mini-theatres in this workshop. The project ended with a final presentation of all mini-theatres.

4   Results and Discussion

After every workshop, changes were made to the design of the mini-theatre. The first mini-theatre was a simple cardboard construction with a servo motor under the stages. Later, two gears were added and different materials were used. Only some minor changes were made to the design of the cardboard stages. Children enjoyed that they could take the stages apart while tinkering.

Our workshops illustrate that the project has the most value when it is organised in multiple sessions over several weeks. In this case, children can reflect on their work during the process of making it. They can still make adjustments before the project is finished. Our interviews with participants show that the Diorama Project works best in a storytelling context. Children can invent their own story, tell a fairy tale or remake a book; the possibilities are endless. Sound recording is a rather unusual activity for schools, but children think it is a great thing to do. While they are using the computer in a purposeful way, they practice speaking skills, empathise with their character and gain self-confidence.

Tinkering the stages of the mini-theatre is by far children's favourite part of the project. It requires teamwork and stimulates spatial thinking, which are both important skills for working in technology and engineering contexts. Crafting supplies, like recycled materials and little household items of all shapes, can encourage creativity.

Observations showed that children struggled with the syntax of Arduino's C-based programming language. To meet the needs of the mini-theatre project, new visual blocks were developed for the visual programming platform *Blockly4Arduino* (Ingegno). As a result, the project stimulates logical thinking without the hassle of forgetting semicolons or brackets.

Literacy is another major component of the project. By inventing a story for their mini-theatre, children enrich their vocabulary, improve writing and communication skills and stimulate their imagination. It encourages pupils to collaborate and share ideas, concepts and experiences. Also, research has shown that girls show more interest in learning computer programming when it is used for the purpose of storytelling (Kelleher and Pausch 2007).

Currently, a gender difference is often noticed in STEM disciplines. Boys are more familiar with construction toys (like LEGO), cars, robots and science topics (Rogers and Portsmore 2004). LEGO is commonly used in engineering classes and many educational tools for programming are robots. As a result, boys feel more confident and courageous when taking part in STEM activities. The mini-theatre concept has another approach; both boys and girls have written and listened to stories before. They also have taken part in tinkering and crafting activities since kindergarten. By doing so, this project strives to be gender equal and stimulate both boys and girls to take part actively during the whole process of making.

Finally, a STEAM project should be fun. The mini-theatre fosters imagination and creativity of children. By storyboarding, sound recording, tinkering and programming, every group will finish with a unique product that reflects their personality and abilities. After all, learning valuable skills and competencies, is the project's goal, but showing a magnificent theatre play is the ultimate goal of children.

5   The Diorama Kit

To organise the workshops a supporting kit was developed. Each team of three to five pupils uses one Diorama Kit, including

- instruction booklets for every activity;
- a microphone for the sound recording activity;
- Microduino mCookie modules;
- all the building parts to assemble the mini-theatre.

The mini-theatre is a modular building kit giving children the freedom to build and rebuild many different constructions for their theatre. All parts of the kit are 3D printed with PLA or laser cut of 3 mm PMMA sheets. Since many materials, such as plywood, MDF and cardboard, are available in 3 mm, pupils and teachers can create extra parts by themselves to enhance their mini-theatres.

The kit is designed in such a way that the four stages can also be replaced by either two or three stages. The stages are not included in the diorama kit. Children can cut them out of cardboard sheets by themselves. Hence, teachers can reuse the diorama kits as many times as they want.

Furthermore, all the production files are open source. Schools, workshop organisers and makers worldwide can use a laser cutter or other production techniques - even a cutter knife and some patience will work - to make the mini-theatre by themselves. They can easily make adjustments or use other electronic modules as well.

Although teachers are free to plan the workshops according to their own schedule, our results show that it is recommended to follow the order given below.

Story writing   starts with a brainstorm about all types of stories that children could tell. Which stories do they enjoy listening to? Have they ever seen a theatre play? Can famous stories be changed? Whatever they can imagine, they can write. Next, each team chooses one story from the brainstorm for their mini-theatre. They learn about storyboarding and the basic elements of a story's structure. Finally, they write the script of their story. The activity takes 1,5 h.

Storyboarding: Before they start tinkering, every team takes a look at their script and storyboard again. Together, they make a plan, so everybody knows what each member is going to make. Ideally, this is done a few days before the tinkering activity, in case pupils want to bring extra crafting supplies for their mini-theatre. The teams can tinker for about two hours, but the stages shouldn't be finished yet after this activity.

Sound recording: This workshop starts by listening to audio clips of theatre plays, animation films, stories and dialogues. Pupils hear that some character have different voices, temper and dialects. This is the art of spoken word and voice acting. It is important that children learn to empathise with their character and use intonation while speaking. After short practice, they can record their play with a voice recording application. This activity should take no longer than 1,5 h.

Tinkering: Since they have already written and recorded their stories, pupils have a better understanding of what they want to make. As they continue tinkering, they can implement new ideas, adjust and improve what they had made, and give feedback to their classmates. They should finish all stages at the end of this workshop (1,5 h) (see Fig. 3).

Fig. 3. A pupil during the tinkering activity

Programming: Everything is ready, now they only have to bring it alive with the electronic modules. First, the children follow the instructions to get familiar with the programming interface. By doing this, they will end up with a basic diorama that can rotate and play sound. Next, they can choose from a list of features to add light, more buttons or extra sounds. By following specific instructions, every group enhances their theatre play with unique features. This activity will take at least three hours, but depending on the interest of the pupils and the time the teacher can give, the workshop can be extended.

Presentation: Each team presents their play to their classmates and teacher. Classmates get the chance to provide useful feedback.

Filming and video editing: After the presentations, this optional workshop can be added. The children can use a camera or phone to film their play. Next, they use a simple video editor, to enrich their story with text, visual and sound effects, or adjust volume and brightness. Ultimately, the video is sent to the diorama platform and shared on the school website.

6   Conclusion and Future Work

The Diorama Project is a series of seven transdisciplinary workshops in which children learn the basics of coding electronics, practise teamwork skills and spatial thinking. Four workshops with nearly 40 children, led to a project that not only supports STEAM disciplines, but many 21st Century skills.

Additionally, it triggers their imagination while they use writing and speaking competencies in a fun and meaningful way. Gender equality is one of the project's strengths, since it starts from storytelling and crafting – activities that both boys and girls are familiar with. *The Diorama Project* is open source available for every educator or maker worldwide. All information for teachers is easily accessible on the Diorama Github[1] page.

While the current version of the project is the result of several workshops with end-users, including qualitative interviews about pupils' experience with using the product, further work could see the testing of skills retention when using the Diorama Project for teaching STEAM.

Acknowledgements.   The authors would like to thank Xi Li and the children who enthusiastically participated in the development of this project.

References

Ananiadou K, Claro M (2009) 21st century skills and competences for new millennium learners in OECD countries. OECD Educ Work Pap 33. doi:10.1787/218525261154

Anderson RE (2008) Implications of the Information and Knowledge Society for Education. In: International Handbook of Information Technology in Primary and Secondary Education. Springer, New York, pp 5–22

Barnard S, Hassan T, Bagilhole B, Dainty A (2012) 'They're not girly girls': an exploration of quantitative and qualitative data on engineering and gender in higher education. Eur J Eng Educ 37:193–204

Beetham H, Sharpe R (2013) Rethinking Pedagogy for a Digital Age: Designing for 21st Century Learning

Blumenfeld PC, Soloway E, Marx RW et al (1991) Motivating project-based learning: sustaining the doing, supporting the learning. Educ Psychol 26:369–398

Dede C (2010) Comparing frameworks for 21st century skills. 21st Century Ski 51–76

Ferrari A, Cachia R, Punie Y (2009) Innovation and Creativity in Education and Training in the EU Member States: Fostering Creative Learning and Supporting Innovative Teaching. Innov Creat E&T EU Memb States 64

Ge X, Ifenthaler D, Spector JM (2015) Emerging Technologies for STEAM Education. Springer, Cham

Gordon J, Halász G, Krawczyk M, et al (2009) Key Competences in Europe: Opening Doors for Lifelong Learners Across the School Curriculum and Teacher Education. CASE Netw Reports. doi:10.1017/CBO9781107415324.004

---

[1] https://github.com/thedioramaproject.

Halász G, Michel A (2011) Key competences in Europe: Interpretation, policy formulation and implementation. Eur J Educ 46:289–306. doi:10.1111/j.1465-3435.2011.01491.x

Ingegno (2017) Blockly4Arduino. http://ingegno.be/01-blockly-4-arduino/. Accessed 30 Apr 2016

Johnson S, Thomas AP (2010) Squishy circuits: a tangible medium for electronics education. In: Proceedings of the 28th of the international conference extended abstracts on human factors in computing systems - CHI EA 2010, p 4099. ACM Press, New York, USA. doi: 10.1145/1753846.1754109

Kelleher C, Pausch R (2007) Using storytelling to motivate programming. Commun ACM 50:58–64. doi:10.1017/CBO9781107415324.004

Lego Education (2017) WeDo 2.0. https://education.lego.com/en-us/elementary/shop/wedo-2. Accessed 30 Apr 2017

littleBits (2017) STEAM Education Class Packs. https://littlebits.cc/kits/steam-education-class-packs. Accessed 30 Apr 2017

Masuda Y (1980) The information society as post-industrial society. World Future Society

Mataric M, Koenig N, Feil-Seifer D (2007) Materials for enabling hands-on robotics and STEM education. AAAI Spring Symp Semant Sci Knowl Integr 99–102

Mcgrath MB, Brown JR (2005) Visual Learning for Science and Engineering. IEEE Comput Graph Appl 25:56–63

Microduino Studio Microduino mCookie: The smallest electronic modules on LEGO®. Kickstarter

Niess ML (2005) Preparing teachers to teach science and mathematics with technology: Developing a technology pedagogical content knowledge. Teach Teach Educ 21:509–523. doi: 10.1016/j.tate.2005.03.006

Peppler K (2013) STEAM-Powered Computing Education: Using E-Textiles to Integrate the Arts and STEM

Reich R (1991) The Work of Nations. Preparing Ourselves for the 21st-Century Capitalism. Vintage Books, New York

Resnick M (2006) Computer as paint brush: technology, play, and the creative society. In: Singer DG, Golinkoff RM, Hirsh-Pasek K (eds) Play = Learning: How play motivates and enhances children's cognitive and social-emotional growth. Oxford University Press, New York, pp 192–208

Rogers C, Portsmore M (2004) Bringing Engineering to Elementary School. J STEM Educ 5:17–29

Roschelle JM, Pea RD, Hoadley CM et al (2000) Changing how and what children learn in school with computer-based technologies. Futur Child 10:76–101

Silva E (2009) Measuring skills for 21st-century learning. Phi Delta Kappan 90:630–634

Tarnoff J (2010) STEM to STEAM – Recognizing the value of creative skills in the competitiveness debate

Thijs A, Fisser P, Hoeven M van der (2014) 21E Eeuwse Vaardigheden in Het Curriculum Van Het Funderend Onderwijs. Slo 128

Tytler R, Osborne J, Williams G, et al (2008) Opening up pathways: Engagement in STEM across the Primary-Secondary school transition

Voogt J, Roblin NP (2012) A comparative analysis of international frameworks for 21st century competences: implications for national curriculum policies. J Curric Stud 44:299–321. doi: 10.1080/00220272.2012.668938

WEF (2015) New Vision for Education Unlocking the Potential of Technology. 1–32

Willingham DT (2010) '21st-Century' skills. Am Educ 17:17–20

Yakman G (2008) STEAM education: An overview of creating a model of integrative education

From Technological Specifications to Beta Version: The Development of the Imprint+ Web App

Pedro Beça[1]([⊠]), Pedro Amado[1], Maria João Antunes[1],
Milene Matos[2], Eduardo Ferreira[2], Armando Alves[2], André Couto[2],
Rafael Marques[2], Rosa Pinho[2], Lísia Lopes[2], João Carvalho[2],
and Carlos Fonseca[2]

[1] DigiMedia, University of Aveiro, Aveiro, Portugal
{pedrobeca, pamado, mariajoao}@ua.pt
[2] CESAM, University of Aveiro, Aveiro, Portugal
{milenamatos, elferreira, armandoalves, andrepcouto,
rafael.a.marques95, rpinho, lisia, cfonseca}@ua.pt,
jlcarvalho@gmail.com

Abstract. Ecologically responsible behaviour and commitment to a sustainable development should be fostered among citizens. The IMPRINT+ project, co-funded by Erasmus+ programme of the European Union, intends to promote an ecological reasoning at an European level, supported by the participation of local communities and particularly by young people. Based on ICT use (smartphones, tablets and websites) and using a gamification strategy, the IMPRINT+ project aims to encourage young citizens to become aware of the global environmental impact of their everyday actions and to encourage them to act and participate in a local context. This paper aims to answer the question of what requirements, functional and non-functional, are appropriate to an app designed to engage young citizens, to be aware of their ecological footprint and to contribute, within different levels (individual, familiar, community or even national), with actions to its compensation. The paper describes the development process of the app, in terms of its specifications, requirements and system architecture. Preliminary tests suggest that the options made to develop the app were appropriate to promote the engagement of young people within the projects' goals.

Keywords: Ecological reasoning · Local community · IMPRINT+ mobile app

1 Introduction

Sustainable development has become a global concern in the last decades (e.g. Rio+20 UN Conference on Sustainable Development; Horizon 2020 agenda; UN Decade of Education for Sustainable Development 2005–2014). Younger European generations have been raised aware of deforestation, global warming, ecological footprint and the demand for preserving the planet for future generations. However, in an increasingly global society, the problem is understood as global and in the need for global solutions. Efforts have been made by school communities, governmental institutions and NGOs

© Springer International Publishing AG 2018
Ó. Mealha et al. (eds.), *Citizen, Territory and Technologies: Smart Learning Contexts and Practices*, Smart Innovation, Systems and Technologies 80, DOI 10.1007/978-3-319-61322-2_18

to promote environmental awareness. Still, European society lacks a broad ecological culture, or a general commitment for solving the 'global problem' through stepwise local actions. Community-led local development (CLLD) actions are at the root of Europe 2020 strategy (European Comission 2014a, b), as a tool for the implementation of the global agenda through the engagement of local actors. However, there is a long way from project to implementation. In a generalized context of economic recession, local leaderships must be able to mobilize available resources, such as natural resources and human capital. Young citizens are one of the strongest changing force in the society but continue to receive contradicting information. Young generations might be aware of their footprint but are not aware how to contribute to its compensation, at an individual, familiar, or community level. They are not truly educated to think and act locally. Therefore it is fundamental to build a new praxis: (i) based on the scientific, technical and empirical knowledge of the territory and its natural and cultural resources; (ii) able to engage the community, and especially young citizens, in the ecological, social and economic enhancement of these resources (Matos et al. 2015).

The goal of IMPRINT+ project is the transnational promotion of an ecological reasoning based on the changing power of local community and on the participation and entrepreneurship of young European citizens. On the one hand, IMPRINT+ aims to build up a strategy for stressing the concept "I have a print" that can be easily compensated, among young school age citizens. The strategy is to focus primarily on the local level of this impact, which is the level where they will be more able to reflect and intervene. Therefore, the logic of compensation will be thought at a local level and will focus on the transformation of the footprint into a positive one—"I have a print+". On the second hand, the project aims to develop a set of outputs that will help young citizens to learn how to estimate the footprint of local impacts and to compensate, locally, for those impacts. The project intends to: (a) produce tutorials and train schoolteachers and local administration technicians on how to approach this concept in order to explore it with their students; (b) produce Information and Communication Technologies (ICT) applications (mobile web app and web platform) to stimulate and support the participation of young citizens. Their engagement is reinforced through the design and implementation of a gamification philosophy. Young citizens could then become a major force of imprinting in society. For that, students should be able to count on the support of their schools and local leaderships. This project targets society as a whole. IMPRINT+ is targeted to younger people (12 to 14 years old – third cycle of primary education; 15 to 18 years old – secondary education), mostly because they are more permeable to new concepts and more familiarized and eager to deal with ICT. On the one hand, they have a greater power for individual action than younger students, who can be more permeable to new concepts, but more dependent on adults for action. On the other hand, they are in a stage where, soon, they will choose their future educational and professional paths. This project can help them decide based on sound knowledge and hands-on experience. The logic of IMPRINT+ is one of local, community-based, intervention for compensating locally caused impacts, but aiming at produce global effects (Matos et al. 2015).

This paper presents a report of an ongoing research project, describing the specification of the integrated mobile web app and web platform associated to IMPRINT+ project (https://imprintplus.org). The next sections are organized as follows: a brief presentation of the state of the art in terms of ICT use to nature conservation and to environmental compensation. Then, a description of the process of the identification of requirements that led to the specification of the ICT solutions of IMPRINT+. We conclude with the presentation of the preliminary results of the field trial.

2 ICT and Nature Conservation/Ecological Compensation

The use of digital technology has been causing major changes in nature conservation. Nature conservation organizations are increasingly using new digital technologies to help achieve their goals (Galán-Díaz et al. 2015; Arts et al. 2015). The adoption of new digital technologies by nature conservation organizations encourages a wide range of changes, for example, in the transmission and processing of data, public involvement, as well as increased digital literacy (Galán-Díaz et al. 2015). To enhance the better assessment and use of these technologies, collaborative initiatives with universities have also been increasing (Galán-Díaz et al. 2015). The collaborative work between institutions enhance opportunities for change in the use of these technologies and generate benefits for organizations and individuals, such as: the possibility of creating new forms of interaction with the public, optimizing work processes, financial benefits and ability to influence policies based on the digital training of institutions. Despite the advantages of using digital technologies, it can also have negative impacts, such as: changing the way people interact with nature, empower the exclusion of certain groups that do not dominate technologies and the difficulty of user involvement, due to the digital barrier (Galán-Díaz et al. 2015).

The use of ICT simplified and facilitated the processes of data transmission and its monitoring or processing. As well as allow for the storage of ever increasing data. It also enabled the emergence of new forms of data collection, for example, through the participation of users, for example via social networks, as well as automatically through software methodologically searching the World Wide Web. The use of mobile devices enables data collection to take place contiguously and in real time, which provides more effective conservation of nature (Arts et al. 2015). Technological evolution, both in terms of the peripherals and functionalities that mobile devices have, for example, camera, GPS, and the appearance of other technologies, such as ibeacons, drones in the IoT domain, allow for new strategies for monitoring and preserving the environment (e.g. the control of illegal logging, poaching, environmental pollution) (Miorandi et al. 2012; Arts et al. 2015). However, as noticed by Arts, van der Wal and Adams (2015), the potential of digital technology to collect and store data presents some problems and raises some questions (Arts et al. 2015).

As a result of the occupation of natural areas, for example, there has been an increased interest in the environmental issues, such as fragmentation and loss of natural resources and in planning methods (Rundcrantz and Skärbäck 2003). Measures to counterbalance the human impact on the environment have often been discussed worldwide, and compensation actions have already been carried out in a systematic way since the beginning of the 20th century.

Concerning ICT use in environmental compensation, locally-based environmental compensation measures are in the heart of other recent projects such as the LIFE+

project "BRIGHT", coordinated by University of Aveiro together with Foundation Mata do Bussaco and Mealhada Municipality; or the project "Futuro – O projecto das 100 mil árvores", coordinated by Catholic University of Oporto and the Administration of the Metropolitan region of Oporto. This reasoning is also at the base of projects like "Limpar Portugal", "Floresta Comum" or "Floresta Unida", promoted by nongovernmental organizations. All these projects involve community (some, specifically, school community) and promote the intervention of local communities in the restoration of habitats and compensation of environmental impacts (Matos et al. 2015).

3 Imprint+ Web App

IMPRINT+ aims to take a step forward, namely, by involving schools and local administration in disseminating and providing technical support to this project, but the aim is to provide enough tools so that young citizens would be able to think, innovate and act on their own, individually or in group, involving their families or at the community level. Sections 3.1 and 3.2 report IMPRINT+ strategy to solve some important projects outputs: (i) development of a mobile web application (web app) for estimating ecological footprint and suggest compensation measures and (ii) development of a web platform (website) with associated database and mapping features.

3.1 Specification, Non-functional and Functional Requirements

As a part of the adopted user-centred development strategy, in order to better achieve the desired outputs, the team modelled the user-requirements based on the target-audience personas and the expected context of use (Lowdermilk 2013). Data needs, product or service qualities, experience attributes and constraints (Sharp et al. 2007; Goodwin 2009) were also taken into account. This information was then used to design the systems' non-functional (NFR) and functional requirements (FR) (Robertson and Robertson 2012). The list contains 17 IMPRINT+ NFR, organized into 8 standard categories (Robertson and Robertson 2012) presented in Table 1. The FR were divided into 18 website requirements and 11 mobile app requirements, organized into the platforms' structural areas and components (Table 2).

The most relevant NFR to highlight were the (1) Look & feel and (2) Usability. Besides security (also a mandatory requirement), the usability is a crucial requirement. Using a subset of usability guidelines for mobile websites and applications (Shitkova et al. 2015), the development team strived for a consistent and usable experience by customizing and optimizing all the graphics and custom illustrations, copy-editing all the texts and ensuring that all the navigation paths were simplified to the bare essentials. The development was done using an agile approach evolving from early digital mockups to test the design, structure and flow (Fig. 1). And develop the second and final interactive digital version (Fig. 2).

While some features, such as modes to support user interaction (communication & feedback) were expectable when designing a social platform (Preece and Maloney-Krichmar 2005; Crumlish and Malone 2009), others were custom designed

Table 1. IMPRINT+ non-functional requirements

#	NFR (website and mobile app)	Categories
1.	Cross platform	Look & feel
2.	Responsive	
3.	English (native language)	Cultural
4.	Easy to new and experienced users	Usability and humanity
5.	Content first. Priority requirements first	
6.	Standard affordances	
7.	Clear navigation	
8.	Consistency between web and mobile interface	
9.	1" or less response time	Performance
10.	Universal login	
11.	User management	
12.	Teenagers' target device (low to mid-end devices)	
13.	Secure data and connections	Security
14.	Latest web standards	
15.	SEO	Operational and environment
16.	Exterior use context	
17.	Single hand interaction	

and developed for the IMPRINT+ web platform. Two are worth noting: (i) the FR12 – an interactive map to search and to create ecological initiatives (interventions or compensations); and (ii) the mobile app FR28 – compensation converter.

The FR12 map uses the Google Maps API, leveraging the target audience familiarity with this solution, and customizes the experience of searching and placing custom flag-like markers of activities. Features such as search by type, and date range filters were implemented in addition to the location.

The FR28 compensation converter is a custom developed solution, that not only calculates the carbon "negative" footprint, but provides a set of "positive" alternatives to compensate the user environmental impact (Fig. 2). These measures range from simpler to more complex ones, and are presented in a contextual flow according to the user input.

Not unique, but still worth noting, is the gamification component. According to the FR13, any user activity on the app will result in earning points, ranking users on the leader boards (FR14). Users can compare and compete in a healthy way, to increase local and international scores. At the same time promoting the use of the app. Gamification enhanced the previous "services with (motivational) affordances in order to invoke gameful experiences and further behavioral outcomes" (Hamari et al. 2014). This not only exposes users to more information by promoting a longer use, but also helps to promote social interaction and a sense of community "by increasing customer value and encouraging value-creating behaviors such as increased consumption, greater loyalty, engagement, or product advocacy" (Hofacker et al. 2016). As the project stipulated, the youngsters have the potential to promote the ecological reasoning on their family and friends circle, in a natural organic social interaction.

Table 2. IMPRINT+ functional requirements (website and mobile app)

#	FR (website)	Categories
1.	General project institutional information	Content
2.	Institutional information about local and international partners	
3.	News	
4.	Documentation	
5.	Browse and search initiatives (interactive, contextual and searchable map)	
6.	Newsletters	
7.	External links	
8.	Free search	
9.	Integrated CMS (front and back-office, with user and content management)	Management
10.	Asynchronous communication mode (registered users)	Social interaction requirements
11.	Ecological carbon footprint calculator (all users)	Specific IMPRINT+ requirements & solutions
12.	Interactive initiatives map (online geographic map)	
13.	Score ranking (registered users)	Gamification components
14.	Interactive leader boards	
15.	Score sharing	
16.	Visit data	Analytics
17.	User actions	
18.	Input source (website or app)	
FR (mobile web app)		
19.	General Project institutional information and links to the website	Content
20.	Intervention areas contextualized information	
21.	Browse and search initiatives (interactive, contextual, searchable map)	
22.	Information area (Info, About, FAQ, etc.)	
23.	Browse and search initiatives (interactive, contextual, searchable map)	
24.	User login/registration	Management
25.	Capture tool (photographic registration and text descriptors)	
26.	Ecological intervention tool	
27.	Carbon print calculator mode	Specific IMPRINT+ Project requirements
28.	Contextual compensation calculator mode	
29.	On-line and off-line syncing of data	Mobile web app specific requirements

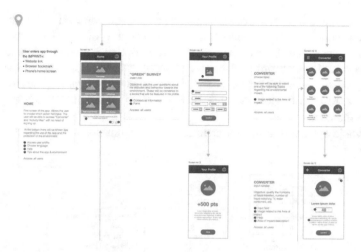

Fig. 1. Early digital mockup depicting the user flow through the mobile web app (detail)

Fig. 2. Interactive version. Converter action example sequence

The requirements were specified according to the priority of development and type user. This allowed to maintain the focus on the unique areas of the project and, eventually, shaped the unique information architecture structure.

3.2 Information Architecture Structure

Due to the overall constraints and benefits, when evaluating a development approach (Charland and Leroux 2011), the team decided early to adopt an integrated rich web app in favour of a native or even cross-platform application (Heitkötter et al. 2013; Xanthopoulos and Xinogalos 2013). This allowed for a more efficient implementation of both the web and mobile interfaces. The mobile browser allows for all the requirements to be fulfilled, ensuring the context and conventions are consistent with users' expectations. Because "[t]he Web technology stack has not achieved the level of performance we can attain with native code" (Charland and Leroux 2011), performance is the only aspect that was considered to be on a compromise: the application is dependant of the mobile browser performance, cache and connectivity. Although some precautions were taken into consideration (e.g. keeping offline data), there are still some issues that would be more transparent by adopting native development. Nevertheless, as mobile platforms and browsers become better performant, the web and mobile apps will respond better. Although there is no guarantee, this future proof design is a result of adopting only web standard solutions, avoiding having to recompile the app.

The whole system information architecture is within the concept of the web platform (Fig. 3).

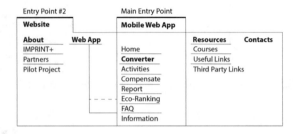

Fig. 3. System information architecture

This is then organized into the two main entry points: (i) the mobile web app; (ii) the website. This allows for a more natural discovery of the project, through the unique features and information the project has to offer. And, in return, we expect for the user to register in a more natural interaction flow and maintain its interest in the project on the long term.

The next section concludes with the findings and discussion of IMPRINT+ web app by presenting the preliminary web app tests and concludes with its main contributions.

4   Findings and Discussion

The web app was first tested on a field trial that occurred in the end of 2016, in Lousada, Portugal. In this field trial, 61 students from three different European countries, with ages from 14 to 17 years old, used the app in a real context. The app was also tested by a group of 25 of the students' teachers. This test mainly identified performance issues in the app (e.g. account synchronization issues and real time geolocalization optimization data). As mentioned before, these were predicted during the development, and will, in time, be minimized as connectivity and device performance grow.

A written system usability questionnaire was also employed in a sample of the participants. Preliminary analysis of the results expressed in the users' answers shows that the app fulfilled its objective by demonstrating a high potential to be used on a daily basis by the target audience, promoting the lifelong ecological reasoning, as its use can enhance the education obtained during the training courses.

As a final note, we would like to conclude by highlight that the web app actively promotes a positive relation between its target adolescent users and the environment. This opportunity to engage this audience is new and unique, as there are no other ecological reasoning mobile applications in this area. And none that specifically implements the compensation and gamification features described.

At the present moment, the IMPRINT+ web app is in use by the target audience users, in informal and formal contexts (such as the training courses promoted by the project partners schools'). Some minor technical problems are being reported and solved. The implementation and evaluation studies will be further explored and presented in future publications about IMPRINT+.

References

Arts K, van der Wal R, Adams WM (2015) Digital technology and the conservation of nature. Ambio 44:661–673. doi:10.1007/s13280-015-0705-1

Charland A, Leroux B (2011) Mobile application development. Commun ACM 54:49–53. doi:10.1145/1941487.1941504

Crumlish C, Malone E (2009) Designing social interfaces: principles, patterns, and practices for improving the user experience. O'Reilly Media, Sebastopol

European Comission (2014a) Guidance on community-led local development for local actors

European Comission (2014b) Cohesion policy 2014–2020 - community-led local development

Galán-Díaz C, Edwards P, Nelson JD, van der Wal R (2015) Digital innovation through partnership between nature conservation organisations and academia: a qualitative impact assessment. Ambio 44:538–549. doi:10.1007/s13280-015-0704-2

Goodwin K (2009) Designing for the digital age. Wiley, Indianapolis

Hamari J, Koivisto J, Sarsa H (2014) Does gamification work? A literature review of empirical studies on gamification. In: 2014 47th Hawaii international conference on system sciences. IEEE, pp 3025–3034

Heitkötter H, Hanschke S, Majchrzak TA (2013) Evaluating cross-platform development approaches for mobile applications. Springer, Berlin, pp 120–138

Hofacker CF, de Ruyter K, Lurie NH et al (2016) Gamification and mobile marketing effectiveness. J Interact Mark 34:25–36. doi:10.1016/j.intmar.2016.03.001

Lowdermilk T (2013) User-centered design: a developer's guide to building user-friendly applications. O'Reilly Media, Sebastopol

Matos M, Ferreira E, Fonseca C (2015) Imprinting an ecological compensation reasoning on society by means of young citizens - IMPRINT+, Erasmus+ Programme application form, Key Action 2: Strategic Partnerships, 80

Miorandi D, Sicari S, De Pellegrini F, Chlamtac I (2012) Internet of things: vision, applications and research challenges. Ad Hoc Netw 10:1497–1516. doi:10.1016/j.adhoc.2012.02.016

Preece J, Maloney-Krichmar D (2005) Online communities: design, theory, and practice. J Comput Commun. http://jcmc.indiana.edu/vol10/issue4/preece.html

Robertson S, Robertson J (2012) Mastering the requirements process: getting requirements right. Pearson Education Inc., Westford

Rundcrantz K, Skärbäck E (2003) Environmental compensation in planning: a review of five different countries with major emphasis on the German system. Eur Environ J Eur Environ Policy 13:204–226. doi:10.1002/eet.324

Sharp H, Rogers Y, Preece J (2007) Interaction design. Wiley, Chichester

Shitkova M, Holler J, Heide T et al (2015) Towards usability guidelines for mobile websites and applications. In: Wirtschaftsinformatik, pp 1603–1617

Xanthopoulos S, Xinogalos S (2013) A comparative analysis of cross-platform development approaches for mobile applications. In: Proceedings of the 6th Balkan conference in informatics on BCI 2013. ACM Press, New York, USA, p 213

Movement Patterns in Educational Games: Comparing A-Priori and Post-Hoc Analyses

Matthias Rehm[✉], Bianca Clavio Christensen, Thorsten Bausbaek Nielsen, Rasmus Albin Rolfsen, and Viktor Schmuck

Technical Faculty of IT and Design, Aalborg University, 9000 Aalborg, Denmark
matthias@create.aau.dk

Abstract. Although movement is essential in location-based games to get from one point of interest to the next, it is seldom taken into account for the game design and the selection of locations. Instead, player movement is usually analyzed after the fact, i.e. when the game is ready to play. In this paper we compare this post-hoc movement analysis with an approach that utilizes the methods for movement analysis to inform the game design itself. We show that both approaches have their merits and solve different tasks, but that there is a benefit of taking movement more serious in designing location-based educational games.

Keywords: Movement analysis · Educational games · Location-based games

1 Introduction

Although movement is one of the major aspects of location-based games, it seldom plays a central role in the development of these games (Paelke et al. 2008) but is often analyzed post-hoc to validate that players reached all game objectives. This contrasts with recent findings that show that game designers would be very interested in incorporating notions of space, direction and movement into location-based games (Brundell et al. 2016). Instead, design and implementation is concentrating on points of interest, which either have a distinguished salience in relation to the aim of the game (e.g. in treasure hunt type games), or are distributed in the environment based on game mechanics (e.g. in tag or capture the flag type games). In the latter case, the real location does not contribute any meaning to the game, making the movement itself the central aspect of the game.

Reid (2008) defines three characteristics of location-based games: (i) size and duration of the game; (ii) infrastructure that is available; (iii) the role that a specific place has in relation to the game. Arguing that the movement is an essential part of a location-based game, we add two more characteristics to this list: (iv) the role movement has in the game; (v) the role the path plays in the game.

Current work has started to look at this role of the path between points of interest and shows for instance how it can become a crucial part of the experience (e.g. Nadarajah et al. 2017). Here, we go a step further and investigate how the game design itself can be informed by an a-priori movement analysis. We compare this approach to a standard Post-hoc analysis that is common in location-based games and show the benefits of both approaches.

© Springer International Publishing AG 2018
Ó. Mealha et al. (eds.), *Citizen, Territory and Technologies: Smart Learning Contexts and Practices*,
Smart Innovation, Systems and Technologies 80, DOI 10.1007/978-3-319-61322-2_19

2 Related Work

Reviewing the literature about location-based (educational) games, it becomes clear that the choice of points of interest in a game is seldom taking the movement between different location into account as an integral part of the experience. Basically, the following features seem to guide the choice of locations: (1) visual salience (e.g. landmarks); (2) coincidence between game and real world (e.g. game-relevant artifact at a heritage site); (3) technical infrastructure (e.g. cell tower); (4) distribution (e.g. evenly distributed across the playing area); (5) random choice.

1. Visual Salience: An approach utilizing visual salience is presented by Lai et al. (2013), where teachers created content for a mobile educational game and could select scenic spots from a map to turn them into points of interest in the game. Another example is given by Rehm et al. (2015) with a mobile game for situated Geometry learning that makes use of landmark buildings near a school to discover and learn about Geometric shapes.
2. Coincidence: Reid (2008) stresses the importance a given place can have in a game and argues that location-based game design should aim for increasing "coincidence" between the real and the game world. She distinguishes between natural coincidence (a natural event (accidentally) happens that is related to the game play), social coincidence (encountering another player involved in the game), and feigned coincidence (deliberate use of props from the physical environment). A similar approach is presented by Kim et al. (2015), where the content authoring includes choosing specific artifacts at a historic sight to become points of interest in the game. Another example is given by Baillie and colleagues (2010), who aim at increasing coincidence between game and real world by choosing locations according to game activities, e.g. the university's health center for recovering.
3. Infrastructure: A different approach is presented by Broll and Benford (2005), where players can collect resources at specific locations and trade them for products at other locations. Resource and trading locations are physically bound to cell towers. Thus, the possible play grounds as well as the possible movements are determined by the physical layout of the mobile infrastructure.
4. Distribution: An interesting twist is presented by Jordan et al. (2013). In the Easter egg hunt game, children must find virtual eggs in a public park. Eggs are distributed evenly across the playing field with some random variation to ensure that the actual location is not predictable. The analysis of player movements is then used to automatically identify landmarks on the playing field.
5. Random Choice: Lund et al. (2010) describe a game architecture, that allows users to design location-based game events for a kind of treasure hunt game including choosing locations for these game events. No principled approach for guiding the user in this task is available resulting in randomly selected locations.

Many studies do not even report the rationale behind the choice of location, although this is a fundamental aspect of the design of a location-based game. Schadenbauer (2009) for instance reports on the design of a mobile treasure hunt in relation to a historic site (defining the area of interaction) but does not discuss, which locations and to which end are incorporated into the game.

Instead of being an integral part of the game design process, the players' movement or potential paths in the real world are usually analyzed post-factum to get an impression over spatial activity in the game, e.g. for distinguishing different player types. Most frequent methods for visualization and analysis of player movement are e.g. heat maps, paths or traversal patterns, allowing for post-hoc analysis of spatial navigation (Reid et al. 2011). These methods usually concentrate on individual players or summarize all movements but lack details about player interactions. Geo-sociograms (Herkenrath et al. 2014) tackle this challenge by displaying distance between players over time to visualize the potential of social interaction during the game.

In the following we are comparing this "classical" approach of post-hoc movement analysis with an approach that aims at identifying movement patterns a-priori to inform the game design.

3   Post-Hoc Movement Analysis: GoPlayDot

3.1   Introduction to GoPlayDot

GoPlayDOT is a game of tag developed for smartphones by GoPlayDot IVS[1]. Users are physically located in the same area such as a public park. The users play in two or three teams, and they are distributed to the different teams before the game starts. The smartphone works as a map, in which a user can navigate in a world of dots. A static dot is a base that is fixed to a specific geographical location, while a moving dot accounts for the position of another user. Each team is represented by different colors to distinguish team members from opponents. The goal is to score the highest amount of points as an individual and as a team before a timer reaches zero. A user gains points by tagging opponent users and bases, which is performed with a single tap on the touch screen. The tagging interaction is rewarded with 1 point for tagging an opponent user and 2 points for a base. A user can only tag a base while being inside it, whereas he can be up to 10 m away from an opponent to perform a successful tag. All bases are unconquered at the start of the game, and a base is scorable as long as it is unconquered or owned by the opponent team. A base can change ownership multiple times during a game, and thus, users can return to a base to collect points. After a user is tagged by an opponent, a user becomes unable to score points, meaning he is dead. To revive, the user must tag a base owned by his team or one of the neutral bases, called a DOThome, which works as a sanctuary in the game and cannot be scored. The playgrounds that the users move within can be over $4000\ m^2$, and therefore, the users have big distances to cover. Danish schools use the game for physical exercises in recess or in gym classes. In other words, the game encourages physical movement to some users, and it was therefore analyzed how the user movement was expressed through the game play and tactics of its users.

---

[1] http://goplaydot.com.

3.2   Movement Analysis

Method(s).   The analysis of GoPlayDOT investigated possible relationships between the score and the user movement in the mobile game. A predicted result was that the users move according to tactics of tagging bases and/or opponents. Like a previous study in competitive games (Pobiedina et al. 2013), the analysis conducted a correlation study to investigate relationships between the observed game metrics, user movement, and user score. The data was obtained from many variables, and therefore, a principal component analysis (PCA) was applied for dimension reduction, meaning that the observed variables were reduced into a smaller number of principal components.

Results.   The data from 10 games with 113 participants in total was collected. The data set only included players who started the game alive or at some point became alive in the game. Hence, players were ignored when they joined the game but never actively participated in the game. Each game had a total game duration of 15 min (corresponding to 900 s). The game duration between participants varied from 402 to 900 s (M: 864.65, SD: 86.49). Data with frequencies below 900 samples are weighted less in the data analysis.

The data is collected for each smartphone joining a game. A data mining tool collected data about the user position, user identification, base positions, and game events. The user positions were collected via the GPS of a user's smartphone, assuming that the smartphone position was equal to the user position. GPS coordinates (latitude and longitude) were collected each millisecond for each smartphone. Each user was differentiated by a unique user id and a team label, indicating to which of the three teams they belonged. Each base was differentiated by their latitude and longitude coordinates. The event metrics were logged when a user pressed the screen, captured a base or an opponent, and received points.

From the dataset, ten features were chosen based on environmental conditions and player interactions:

1. Opponents: The number of opponents that a participant has.
2. Team size: The number of participants on the team that the user belongs to.
3. Playground shape: The ratio between the width and the height of a playground calculated in meters.
4. Bases: The number of bases on a playground.
5. Visited bases: The number of different bases that the user has entered.
6. Base tags: The number of times that a user tags a base.
7. Opponent tags: The number of times that a user tags an opponent.
8. Group: A grouping of two types of user movement.
9. Base distance: The distance between a user and the base closest to him in meters.
10. Opponent distance: The distance between a user and the opponent closest to him in meters.

The first four features were calculated within a team in a game, and these described some preconditions of the game, such as the distribution of the teams and the size of a playground. There are no limits to the number of participants in a team, and the size of a playground is not fixed. Therefore, the distribution of the players and the amount of space that they can move in might have an influence on the user interactions in a game. The last six features described player interactions, and these were calculated for each user. These features were based on previous studies investigating game outcome (Pobiedina et al. 2013), distance between players (Drachen et al. 2014), and path styles (Sookhanaphibarn and Thawonmas 2009). The features of the game outcome included the number of times a player tags bases and opponents during a game, as the user score was an aggregation of these two game metrics. User movement was measured by calculating the distance to the nearest base and opponent from each user. In other words, these two features indicated how close the user is to scoring a point. The last two features described how a player used the space of a playground, and if that could be categorized as different path styles. In analyzing the user trajectories, two path styles were found. The first group was characterized as moving back and forth between bases and repeating their movement. They often moved back to a previous base and stayed within the same area. This movement was similar to a role called guards by Chittaro and Ieronutti (2004). The second group was characterized as moving to new bases rather than the previous ones. Consequently, they moved more and followed the bases in a line or in a circle. This movement was similar to a role called patrols in by Chittaro and Ieronutti (2004). Figure 1 shows examples for these movement patterns from three different playing fields.

Fig. 1. Movement patterns for guards (above) and patrols (below) on three different playing fields (left, middle, right).

The ten features computed an adequate Kaiser-Meyer-Olkin value (KMO = 0.5) according to Field (2000), meaning that the data is likely to factor efficiently. From the PCA a few principal components were retained for further analysis. Based on the Kaiser rule, the components with eigenvalues greater than 1.0 can be retained (Field 2000). For this data set, three components were retained, which account for 61.13 percentage of the variance in the data set.

The component loadings showed that the first principal component was correlated to visited bases, base tags, group, opponent distance, base distance, and opponent tags. The last feature had a negative correlation to the component. These six features described how a user score points and moves according to opponents and bases, and therefore, this component was named Tactics. The loadings for this component can be interpreted by looking at the two different groupings of users, where the patrols account for the positive correlation in the component, and the guards account for the negative one. The second component was correlated to playground shape, team size, bases, and opponents. These features described the environmental characteristics, and the values for each feature were the same within a team. This component was named Game Conditions, as the features described some pre-conditions of a game rather than the actual user movement and performance during a game. The third component was positively correlated to visited bases and bases, while it was negatively correlated with base distance, playground shape, and team size. The interpretation of this component can be when there are many bases on a playground: a user visited more bases, and a user was naturally closer to a base. Furthermore, a user's distance to a base was reduced for some playgrounds, and when there were few members in a group. Ultimately, this component was named Interaction With Bases, where the interaction refers to base distance and visited bases.

The most interesting result was the correlations between the first component and the features. It suggested that the patrols visited and tagged more bases, and oppositely, the guards tagged more opponents instead of visiting and tagging bases. Furthermore, the guards were closer to a base and an opponent during a game, in comparison to the patrols. Naturally, when the guards were averagely closer to an opponent, they had the chance to tag one more frequently. Furthermore, they might have stayed near a base for a longer period, in comparison to patrols, as they were averagely closer to a base and visited fewer bases.

The data set contained a sample of the existing users of GoPlayDOT, and therefore, there might be additional tactics applied by other users of the mobile game. Additionally, the last component suggested additional tactics or movement interactions, since it was unlike the patrol and the guard tactics. The component also suggested that there might be combinations of the patrol and the guard tactics. An explanation can be that the users had the freedom to move everywhere on the playground, and therefore, they also had the freedom to create individual tactics or change tactics during the game. Thus, a user's movement was unique and included randomness. It was therefore a difficult task to identify the movement interactions of an entire game. For future work, it could be interesting to analyses at what event or time in the game, a user applied one tactic and when he changed tactics. This could perhaps also reveal why the user moved in a specific pattern.

3.3 Features of Post-Hoc Analysis

We have presented a "classical" movement analysis for a mobile game. Points of interest in the game are based on distribution across the area to facilitate movement between locations. The real-world features of locations have no effect on the game play. The analysis focused on movement patterns in the area, making use of path visualization. A result is the identification of two distinct game tactics, which are then linked to game play features through principal component analysis.

These (descriptive) results allow for generating specific hypotheses concerning e.g. the game play, placement of locations, or motivational factors for different player types. Thus, such a post-hoc analysis can inform more concrete experimental studies as well as the further development of the game. Because one aim of the game is to encourage movement for school children, a reasonable modification could be to discourage the guard tactics and encourage tactics that lead to more exploration.

4   A-Priori Movement Analysis: Run&Learn Astronomy

4.1   Introduction to Run&Learn Astronomy

The project develops an experiential educational game about the solar system that features space travel by actual physical movement between planets. The aim of the study presented here was to investigate potential movement patterns of fourth graders (10–11 years of age) when they play such a game. The analysis of these movement patterns will feed back into the game design, which will be done by the pupils themselves. For this study, a prototype was developed as a tool to evaluate the influence of different spatial setups for the planets. The prototype was used on the sports field of the school, where participants had the task to find and visit five planets in limited amount of time (see Fig. 2). On reaching a planet, they had the possibility for a short interaction revealing some information about the planet.

Fig. 2. Linear (left) and scattered (right) test setup. The blue marker highlights the starting position of participants and the red markers highlight the position of planets. S - Start, 1 - Sun, 2 - Mercury, 3 - Venus, 4 - Earth, 5 - Mars.

4.2   Movement Analysis

Method(s). By drawing in experience from related work, various movement pattern visualization methods were utilized: Trajectory, vector, network and flow maps based on an account by Orellana et al. (2009); heat maps based on account by Drachen and Canossa (2009); and space-time cubes based on an analysis by Orellana et al. (2012). These methods defined some of the independent variables that had to be logged during the evaluation scenario like GPS coordinates and time stamps.

The project was carried out in collaboration with a local school, where three fourth grade classes participated. Thus, the test was run with 60 pupils of age 10 to 11. Their task was to find 5 planets placed in the designated area using the prototype application.

The test ended when they had either found and interacted with the five planets, or if a time limit of 15 min was reached. Each participant had an individual tablet, and at each time, four participants where exploring the playing field, starting with an offset of approximately 30 s. The test had two different spatial setups, a scattered version where the planets were scattered around the sports field, and a linear version where the planets were placed along a line. The starting position of the participants with the positions of the celestial bodies can be seen in Fig. 2 (left) for the linear and in Fig. 2 (right) for the scattered spatial setups.

To analyze how the spatial setups affected the movement of participants, the following visualization methods were used: trajectory plots, heat maps, and space time cubes. In addition, the distance participants travelled, the time to complete the task, and the number of planets the participant interacted with was calculated.

Results. The type of statistical analysis performed on the data is descriptive, revealing the different movement patterns that are associated with the variation in the spatial layout. By analyzing the trajectories, heat maps and space-time cubes the following results were reached. The trajectories indicate that the linear setup generates less but more radical turns and the pupils generally had an easier time staying on the linear route, with a small deviation between the different paths that the participants took. The scattered setup generated a lot of tight turns, with fewer sections where participants walked straight and hence, produced more complicated patterns. The scattered setup also generated more unique paths and routes between the participants. Examples of the trajectories from the linear and scattered setup can be seen in Fig. 3. The heat map shows a lot of presence along the linear route for the linear setup (Fig. 4, left) and in general, both setups show strong heat signatures at the proximity of the planets. The heat map for the scattered setup also shows heat signatures where participants lost their way and generally shows presence in a larger area of the sports field (Fig. 4, right). The space-time cubes were used to represent the order in which participants visited different planets. They also show when and where a participant would spend a certain amount of time at a certain position (Fig. 5).

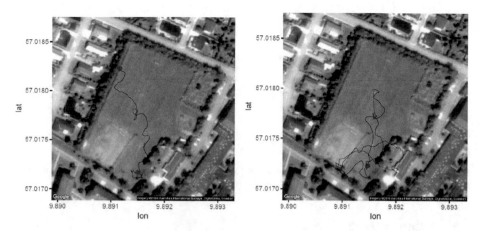

Fig. 3. Example trajectories for linear (left) and scattered (right) spatial setup. The bottom red dot indicates the start, the upper red dot indicates the end of the trajectory.

By using the logged times, coordinates and interaction with planets, the following results were calculated. The participants using the scattered setup travelled a longer distance. However, the participants using the linear setup spent less time finding the planets and found more planets overall.

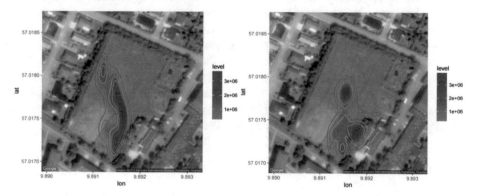

Fig. 4. Example heat maps for linear (left) and scattered (right) spatial setup. The warmer the colour the more time the participant spent in the position (measured in number of location samples taken).

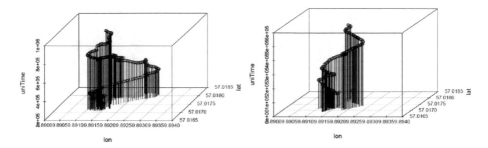

Fig. 5. Example of space-time cube for linear (left) and scattered (right) setup. The space-time cube adds the time dimension to the trajectory information.

4.3   Features of A-Priori Analysis

In this section, we presented a study that aimed at utilizing the analysis of movement dynamics to inform an important aspect of the design of a location-based game, i.e. which locations to select for interactions. Whereas this question is easily answered for specific types of games, usually games that aim at conveying information about specific locations (e.g. Nadarajah et al. 2017), it is not trivial for games where the coincidence between game and real world is not a 1:1 relation. This is the case for the Astronomy game that is under development here and thus, it was necessary to find a robust way for deciding how and where to place points of interest in the game area.

Although the scattered spatial layout is more realistic in depicting the relative position of planets towards each other and furthers more exploration, results show that it might hinder reaching the learning goals set by the curriculum like gaining knowledge about the distances between planets and information about the planets themselves. This is easier achieved with the movement that is induced by the linear layout, which makes comparison of the distance between planets easier and guarantees that players will actually find planets to gain information about them.

5   Conclusion

To conclude, we have shown the merits of post-hoc and a-priori analysis of movement patterns of players in location-based games. While post-hoc tests allow for analyzing spatial behavior in depth based on a given game design, a-priori analysis feeds directly into the game design itself and presents a tool for creating meaningful points of interests and paths between these locations.

Acknowledgments. We would like to thank the pupils and teachers at Sofiendalskole in Aalborg for their kind cooperation in the Run&Learn project.

References

Baillie L, Morton L, Uzor S, Moatt DC (2010) An investigation of user responses to specifically designed activities in a multimodal location based game. J Multimodal User Interfaces 3(3): 179–188. doi:10.1007/s12193-010-0039-z

Broll G, Benford S (2005) Seamful design for location-based mobile games. Springer, Berlin, pp 155–166. http://dx.doi.org/10.1007/11558651_16

Brundell P, Koleva B, Wetzel R (2016) Supporting the design of location-based experience by creative individuals. In: Proceedings of the 11th international workshop on semantic and social media adaptation and personalization (SMAP)

Chittaro L, Ieronutti L (2004) A visual tool for tracing users' behavior in virtual environments. In: Proceedings of the working conference on advanced visual interfaces, AVI 2004. ACM, New York, pp 40–47. http://doi.acm.org/10.1145/989863.989868

Drachen A, Canossa A (2009) Analyzing spatial user behavior in computer games using geographic information systems. ACM, Finland, pp 182–189

Drachen A, Yancey M, Maguire J, Chu D, Wang IY, Mahlmann T, Schubert M, Klabjan D (2014) Skill-based differences in spatio-temporal team behavior in defence of the ancients 2. In: Proceedings of games media entertainment (GEM). IEEE

Field A (2000) Discovering statistics using SPSS for windows: advanced techniques for beginners. Sage Publications, Thousand Oaks

Herkenrath G, Heller F, Huch C, Borchers J (2014) Geo-sociograms: a method to analyze movement patterns and characterize tasks in location-based multiplayer games. In: Proceedings of CHI 2014. ACM, New York, pp 1501–1506

Jordan KO, Sheptykin I, Grueter B, Vatterrot HR (2013) Identification of structural landmarks in a park using movement data collected in a location-based game. In: Proceedings of the first ACM SIGSPATIAL international workshop on computational models of place, COMP 2013. ACM, New York, pp 1:1–1:8. http://doi.acm.org/10.1145/.2534853

Kim H, An S, Keum S, Woo W (2015) H-Treasure hunt: a location and object-based serious game for cultural heritage learning at a historic site. Springer, Cham, pp 561–572. http://dx.doi.org/10.1007/978-3-319-20609-7_53

Lai HC, Chang CY, Li WS, Fan YL, Wu YT (2013) The implementation of mobile learning in outdoor education: application of QR codes. Br J Educ Technol 44(2):57–62

Lund K, Lochrie M, Coulton P (2010) Enabling emergent behaviour in location based games. In: Proceedings of the 14th international academic mindtrek conference: envisioning future media environments, MindTrek 2010. ACM, New York, pp 78–85. http://doi.acm.org/10.1145/1930488.1930505

Nadarajah S, Overgaard B, Pedersen P, Rehm M, Schnatterbeck C (2017) Enriching location-based games with navigational game activities. Springer, Heidelberg

Orellana D, Bregt AK, Ligtenberg A, Wachowicz M (2012) Exploring visitor movement patterns in natural recreational areas. Tour Manag 33(3):672–682

Orellana D, Wachowicz W, Andrienko N, Andrienko G (2009) Uncovering interaction patterns in mobile outdoor gaming. IEEE, pp 177–182

Paelke V, Oppermann L, Reimann C (2008) Mobile location-based gaming. Springer, Berlin, pp 310–334. http://dx.doi.org/10.1007/978-3-540-37110-6_15

Pobiedina N, Neidhardt J, Calatrava Moreno, MdC, Werthner H (2013) Ranking factors of team success. In: Proceedings of the 22nd international conference on world wide web, WWW 2013 Companion. ACM, New York, pp 1185–1194. http://doi.acm.org/10.1145/2487788.2488147

Rehm M, Stan C, Wøldike NP, Vasilarou D (2015) Towards smart city learning: a van hiele inspired location-aware game. In: International conference on entertainment computing (ICEC). Springer, pp 399–406

Reid J (2008) Design for coincidence: incorporating real world artifacts in location based games. In: Proceedings of the 3rd international conference on digital interactive media in entertainment and arts, DIMEA 2008. ACM, New York, pp 18–25. http://doi.acm.org/10.1145/1413634.1413643

Reid J, Hull R, Clayton B, Melamed T, Stenton P (2011) A research methodology for evaluating location aware experiences. Pers Ubiquit Comput 15(1):53–60. doi:10.1007/s00779-010-0308-6

Schadenbauer S (2009) Mobile game based learning: designing a mobile location based game. Vieweg+Teubner, Wiesbaden, pp 73–88. http://dx.doi.org/10.1007/978-3-8348-9313-0_6

Sookhanaphibarn K, Thawonmas R (2009) A movement data analysis and synthesis tool for museum visitors' behaviors. Springer, Berlin, pp 144–154. http://dx.doi.org/10.1007/978-3-642-10467-1_12

Automating Assessment of Exercises as Means to Decrease MOOC Teachers' Efforts

Vincenzo Del Fatto[✉], Gabriella Dodero, Rosella Gennari, Benjamin Gruber,
Sven Helmer, and Guerriero Raimato

Faculty of Computer Science, Free University of Bozen-Bolzano,
Piazza Domenicani 3, 39100 Bolzano, Italy
{vincenzo.delfatto,gabriella.dodero,guerriero.raimato}@unibz.it,
{gennari,shelmer}@inf.unibz.it,
benjamin.gruber@stud-inf.unibz.it

Abstract. To increase digital fluency in adult population, efforts have included offering e-learning initiatives and MOOCs, and more efforts should be undertaken in order to facilitate MOOC implementation and to widen MOOC participation. Along this line, this paper presents the first findings toward the automated assessment of bash scripting exercises, to be offered in a MOOC focused on this topic. By using exercise solutions submitted by participants during a past edition of such a MOOC, we implemented bash scripts able to semi-automatically assess student submissions. Tests on three different exercises showed a decrease of 50%, w.r.t. actual manual assessment time, measured while the MOOC was delivered.

1 Introduction

The importance of a good level of initial and continuing education among citizens of a territory has been recognized in literature on smart territories (Giovannella 2015). Such information is used by governments to lead educational investments to improve in areas such as digital literacy of adult citizens. To aim at digital fluency, efforts have been undertaken to support adults' learning by means of e-learning initiatives and Massive Open On-line Courses (MOOCs), in parallel with school based initiatives reaching the younger population. In line with the above considerations, in the past years, our research group has developed Open Educational Resources (OERs) for high schools, and has recently offered a MOOC, open to all citizens and welcome by high school teachers, where digital fluency with Open Source technologies was the main educational goal. The choice of open source technologies as target for both the OERs, and the MOOC built upon them, is based on strategic considerations that the authors are sharing, which have been expressed by various members of the Free Open Source Software (FOSS).

As an example, the use of FOSS in high schools, which are heavily supported by taxpayer's money, would, on one hand, save a significant amount of public money, that today is spent for paying proprietary licenses; on the other hand, it would allow to "open up" the technology itself (see many educational examples in (FOSS 2017)), this way moving naturally from digital literacy (how and what) to digital fluency (when and why).

© Springer International Publishing AG 2018
Ó. Mealha et al. (eds.), *Citizen, Territory and Technologies: Smart Learning Contexts and Practices*,
Smart Innovation, Systems and Technologies 80, DOI 10.1007/978-3-319-61322-2_20

Using FOSS when teaching digital literacy or digital fluency today has to "fight" against misconceptions about it being less effective or less user friendly than commercial software. In our region, the long lasting success of the FUSS project (FUSS 2017) at school has however lowered such barriers, and created the background for a wider participation in the MOOC itself.

In details, our MOOC, called E3OSMOOC, has been focused on a specific FOSS technology, bash scripting, with the aim to make all learners aware of what happens "inside" their computer, beyond the usual GUI, and able to manipulate files and directories as bash "objects", as system administrators would do. A secondary objective to be achieved with our MOOC is to "teach by examples" good programming principles, such as checking values of input parameters before using them, which are unfortunately not stressed enough in many introductory programming courses, and cause so many bugs in beginner's code. In particular, a new teaching methodology called eXtreme Apprenticeship (XA) was integrated into the MOOC. Our experiences with applying XA to teaching bash scripting, both at University and at High School level, showed an optimal ratio between teachers and students, so that teachers can manually assess all exercises done by students. Considering the typical high number of participants to a MOOC, manual assessment of exercises could represent a problem. In order to widen future participation on similar MOOCs on bash scripting, we recently developed a strategy, to semi-automatically assess bash scripting exercises.

The paper is organized as follows. Section 2 presents the state of the art. Section 3 describes the characteristics of the E3OSMOOC. Section 4 describes our first findings in the automated assessment of bash scripting, which were applied to the data collected from E3OSMOOC delivery. Section 5 concludes the paper with final considerations and ideas for future work.

2   State of the Art

The E3OSMOOC is the outcome of six years of application of the XA learning methodology at the Free University of Bozen-Bolzano. Bash programming activities in Operating System Lab activities consisted on 52 exercises, to be done in six weeks. Although the topic is usually considered tedious and difficult by students, XA has proven useful to encourage students to overcome their difficulties in bash programming (Del Fatto et al. 2014a, b, 2016a). The first 30 of such exercises became comic strip videos, by using aesthetics and gamification principles (Del Fatto et al. 2015c), and were first made available as OERs to high school teachers, then used as support material in the E3OSMOOC. XA is a comprehensive approach for organizing education in formal contexts, based on Cognitive Apprenticeship (Collins 2006): a new task is learned by apprentices, looking at the master performing it, and then repeating it under master´s guidance. XA has been applied to teaching new cognitive skills, at BSc level, e.g., Vihavainen et al. (2011). Results achieved so far are impressive: reduction of dropout rates, higher grades, and higher retention of learned topics. In Vihavainen et al. (2011) this is ascribed to XA increasing learning performances of average and below average students, who would otherwise fail in traditionally taught courses. Italian school teachers experienced with the XA methodology on various science related topics (Del Fatto et al. 2015a, b).

Although there is work (Douce et al. 2005; Ihantola et al. 2010; Pieterse 2013; Spacco et al. 2006) about automated assessment of exercises, as well as quiz-like games in adaptive learning systems, e.g. Di Mascio et al. (2012), we found only one experience on automated assessment of XA exercises. Pärtel et al. (2013) proposed the Test My Code (TMC) system that supports the instructor to manage XA exercises and their assessment, which was successfully used in MOOCs on programming, and in university courses. Note that TMC manages type safe high-level programming languages, such as Java, and in our case, features of the bash interpreter prevented us from using TMC.

3   E3OSMOOC

This Section describes the main characteristics of E3OSMOOC, the source of data we used to apply semi-automated assessment of exercises. In particular, it highlights in Sect. 3.1 how it was organized, in Sect. 3.2 how its learning material was conceived, and in Sect. 3.3 which gamification elements were introduced to motivate MOOC participation, where gamification is intended as the choice of game design elements to engage different users, as in Gennari et al. (2017). E3OSMOOC was delivered from March 1, 2016 to April 30, 2016 (Del Fatto et al. 2016b).

3.1   MOOC Organization

E3OSMOOC followed XA principles, providing 30 bash programming exercises, which required learners to progressively build their skills. The bash environment is usually considered more difficult to master, with respect to programming environments used in teaching other languages, e.g. Java. It however represents a typical BSc course topic at intermediate level, which is further deepened only by professionals working as system administrators. E3OSMOOC was implemented on a Moodle platform, and organized into progressive modules, each one based on a specific learning objective:

1. Being able to use the Command Line Interface in Linux. Ex. 1–7.
2. Being able to create bash scripts as sequences of commands, with variables and formatted messages. Ex. 8–14.
3. Being able to create scripts with parameters. Ex. 15, 16, 17.
4. Being able to create scripts that collect and filter information from text files, with the use of pipes and variables. Ex. 18–22.
5. Being able to test script parameters validity, existence of files and directories, perform regression testing and incremental development of scripts. Ex. 23–30.

Exercises were assessed as passed or not passed, with the possibility to re-submit the same exercise until pass, as many times as needed. Following the XA approach, students received a formative assessment as feedback from the course teachers. Failed exercises always received a short comment concerning mistakes, while feedback for passed exercises was often an encouragement.

3.2   MOOC Learning Material

For each exercise, MOOC participants had available: a comic strip video, explaining step by step how to do it; the video transcript in three languages; and one or more assignments, to upload exercise solutions. To provide multi modal access to learning material, video-lessons were designed considering usability principles (Nielsen et al. 1990), and cartoon-like characters made them more eye-catching (Del Fatto et al. 2014c). Each video was realized in a specific language, and its transcripts were always provided in three languages, in order to avoid de-motivating students who were not fluent in the language of the video. Text was also enriched with visual information (different color background) to guide student towards solutions.

3.3   Gamification

In our experience, bash scripting has always been considered complex and tedious by students. To counter this attitude and increase motivation, the MOOC provided three different gamification elements. The first element, a progress bar, displays progress in the educational path, showing completed correct exercises in green, and failed exercises in red.

The second element, Open Badges, marks the completion of each learning module by awarding a prize. Each module was turned into a milestone to reach, and awarded with an Open Badge to show improvement in skills, to counter the perceived difficulty of solving all 30 exercises before obtaining any recognition.

The third element, a cycling race in stage metaphor, was intended to stimulate a mild competition among MOOC participants. Each exercise submission represents a stage in the race, and exercise delivery time is the stage completion time. Having classified exercises into two different difficulty levels, easy and hard, they correspond in the cycling metaphor respectively to flat stages or mountain stages. Rankings and Reports were the last gamification element. Participants saw their respective position, in stages and in general classification: we noticed that the most active participants actually started competing against one another towards faster exercise delivery.

4   Toward the Automated Assessment of Bash Scripts

In our experiences in applying XA to bash scripting, both at University and High School level (Del Fatto et al. 2015a, b), a possible drawback was identified, which is related to the high number of exercises the teacher has to manually assess. An optimal ratio is estimated to be about one teacher every ten students (Del Fatto et al. 2015a). In E3OSMOOC, solutions of the 30 exercises, submitted by 56 participants, were manually assessed by 5 teachers, for a total of 1500 assessed exercises, over a time period of two months. Again, the ratio one teacher every ten students had been met. However, to allow a wider participation in future editions of the MOOC, we needed an automated (or semi-automated) assessment of bash scripts. We identified a subset of exercises which could be suited to semi-automated assessment, and we implemented some 'correction scripts', that checked if the submitted exercise solutions were correct, by comparing their outputs with given master sample solutions. By using such 'correction scripts', the teacher saves on assessment time of exercises that are automatically marked as correct, and can concentrate on providing feedback to exercises that are marked as wrong. In the following we describe tests of the 'correction scripts' applied to three MOOC exercise solutions. In order to avoid timing differences due to different assessment methods employed by teachers, three exercises were identified, that had been assessed by the same teacher, namely Exercises 1, 7 and 19. Exercises 1 and 7 were classified in the MOOC as easy, while Exercise 19 was classified as hard.

Table 1 shows for each identified exercise, the number of solutions submitted by MOOC participants, the number of correct (T-Correct) and incorrect (T-Incorrect) solutions identified by the teacher during MOOC delivery, the number of correct (A-Correct) and incorrect (A-Incorrect) solutions based on the execution of the 'correction scripts', and the percentage of needed manual checks by the teachers after semi-automated assessment. Note that for Exercises 1 and 19 the 'correction scripts' obtained the same result of the teacher assessment, in terms of numbers of identified correct and incorrect solutions. For Exercise 7, the 'correction script' identified five potentially incorrect solutions, while the teacher identified only one incorrect solution.

Table 1. Results in terms of teacher assessment and automated assessment.

Exercise	Submitted solutions	T-Correct	T-Incorrect	A-Correct	A-Incorrect	Check needed
1	57	56	1	56	1	1.7%
7	44	43	1	39	5	11%
19	41	28	13	28	13	32%

Figure 1 shows (red) actual" teacher's assessment time (in seconds) of Exercises 1, 7 and 19, computed during MOOC delivery. Blue bars show assessment time computed in our test, by running the 'correction scripts' on those same solutions submitted by MOOC participants. Note that automated assessment time increases with the number of incorrect solutions found, that is, when the teacher has to manually correct all solutions marked as incorrect by the 'correction script', to give a specific feedback to each participant. Assessment time takes into account:

- The time, 5 s, to download the solution from the platform, which is constant for all exercises,
- The time needed to assess the exercise, which depends on the specific exercise for manual assessment, and is constant for automated assessment,
- The time, 15 s, to record the grade, which is constant for all exercises.

In particular, for Exercise 1 the teacher spent 2352 s (~39 min) for assessing 57 exercise solutions, while a teacher with the 'correction script' would have spent 1252 s (~21 min). For Exercise 7, the teacher spent 3780 s (~63 min) for assessing 44 solutions, while a teacher with the 'correction script' would have spent 1052 s (~18 min). Finally, for Exercise 19 the teacher spent 5544 s (~92 min) for assessing 41 solutions, while a teacher with the 'correction script' would have spent 2464 s (~41 min). In total, time that would have been saved in assessing Exercises 1, 7 and 19, was 50% of the time

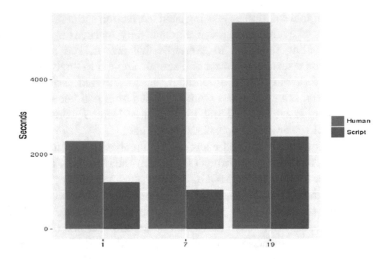

Fig. 1. Automated assessment time vs. Teacher assessment time.

spent by the teacher in assessing such exercises during MOOC delivery. However, by taking into account only the time to execute exercise testing, and not considering the time to download the solution, and the time to record the grade, better results are scored. In particular, Table 2 shows that in case of manual assessment, T-testing is predominant, while in case of automated assessment, A-testing is less than 10% of the total time. These good results motivated us to continue the investigation in that direction, with a MSc thesis just completed, dealing with automated bash script assessment. Details on the 'correction script' can be found in Gruber (2017).

Table 2. Assessment time for the exercises, divided into download, testing and grading time.

Exercise	T-Download	T-Testing	T-Grade	A-Download	A-testing	A-Grade
1	12%	52%	36%	23%	9%	68%
7	6%	82%	12%	23%	9%	68%
19	4%	89%	7%	23%	9%	68%

5   Conclusion

This paper presented the first findings in semi-automatic assessment of bash scripting exercises, following the XA learning methodology. Automating assessment of XA exercises would be important to widen participation in future MOOCs on bash scripting, without increasing teachers number beyond what would be economically affordable. To assess the feasibility of semi-automated assessment, we used solutions proposed by participants during a MOOC delivered in 2016, and implemented some 'correction scripts' which the teacher can use in order to save a significant fraction of assessment time. Such 'correction scripts' can automatically mark exercises as correct, so that the teacher can concentrate on assessing exercises that are marked as wrong, and give a better feedback to help students in understanding their own mistakes. First experiments on the data collected from our MOOC, by semi-automatically assessing three exercises out of 30, showed good results in terms of assessment time decrease. In future, we plan to implement 'correction scripts' for all the exercises, devising more general solutions applicable to different types of exercises, with the final aim to at least halve manual assessment time during MOOC delivery. Such figures would make it feasible to launch periodically new E3OSMOOC editions, providing high quality support to beginners, as per XA methodology.

References

Collins A (2006) Cognitive apprenticeship. In: Sawyer RK (ed) The cambridge handbook of the learning sciences. Cambridge University Press, New York

Del Fatto V, Dodero G, Gennari R (2014a) Assessing student perception of extreme apprenticeship for operating systems. In: Proceedings of the 2014 IEEE 14th international conference on advanced learning technologies, ICALT 2014. IEEE Computer Society

Del Fatto V, Dodero G, Gennari R (2014b) Operating systems with blended extreme apprenticeship: what are students' perceptions? IxD&A 23:24–37

Del Fatto V, Dodero G, Gennari R (2016a) How measuring student performances allows for measuring blended extreme apprenticeship for learning bash programming. Comput Hum Behav 55(PB):1231–1240

Del Fatto V, Dodero G, Lena R (2015a) Experiencing a new method in teaching databases using blended extreme apprenticeship. DMS 2015:124–130

Del Fatto V, Dodero G, Raimato G (2016b) All about the Mooc, minute by minute. Mondo Digitale 15(64). (in Italian)

Del Fatto V, Dodero G, Gennari R, Mastachi N (2014c) Extreme apprenticeship meets playful design at operating systems labs: a case study. Springer, Heidelberg

Del Fatto V, Barazzuol B, Bolliri C, Ferrante I, Osti S, Serra Z (2015b) Extreme apprenticeship really works. Mondo Digitale 14(58):6–11. (in Italian)

Di Mascio T, Gennari R, Melonio A, Vittorini P (2012) The user classes building process in a TEL project. In: Advances in intelligent and soft computing, vol 152. AISC, pp 107–114

Douce C, Livingstone D, Orwell J (2005) Automatic test-based assessment of programming: a review. J Educ Resour Comput 5(3):4

FOSS sourcebook. https://project.inria.fr/foss-sourcebook/. Accessed Mar 2017

FUSS project. https://www.fuss.bz.it/. Accessed Mar 2017

Gennari R, Melonio A, Raccanello D, Brondino M, Dodero G, Pasini M, Torello S (2017) Children's emotions and quality of products in participatory game design. Int J Hum Comput Stud 101:45–61

Giovannella C (2015) Territorial smartness and the relevance of the learning ecosystems. In: 2015 IEEE first international smart cities conference (ISC2), pp 1–5

Gruber B (2017) Semi-automatic recognition of script-based errors in a closed MOOC Environment. MSc in Computer Science, Free University of Bozen-Bolzano

Ihantola P, Ahoniemi T, Karavirta V, Seppälä O (2010) Review of recent systems for automatic assessment of programming assignments. In: Proceedings of the 10th Koli calling international conference on computing education research. ACM, New York

Nielsen J, Molich R (1990) Heuristic evaluation of user interfaces. In: Proceedings of the SIGCHI conference on human factors in computing systems, CHI 1990. ACM, New York

Pärtel M, Luukkainen M, Vihavainen A, Vikberg T (2013) Testmy code. Int J Technol Enhanc Learn 5(3/4):271–283

Pieterse V (2013) Automated assessment of programming assignments. In: Proceedings of the 3rd computer science education research conference, CSERC 2013. Open Universiteit, Heerlen, pp 4:45–4:56

Spacco J, Hovemeyer D, Pugh W, Emad F, Hollingsworth JK, Padua-Perez N (2006) Experiences with marmoset: designing and using an advanced submission and testing system for programming courses. SIGCSE Bull 38(3):13–17

Vihavainen A, Paksula M, Luukkainen M (2011) Extreme apprenticeship method in teaching programming for beginners. In: Proceedings of the 42nd ACM technical symposium on computer science education, SIGCSE 2011. ACM, New York, pp 93–98

The ebook as a Business Communication Strategy

Liliana Cecilio[1(✉)], Dora Simões[1], and João Carapinha[2]

[1] University of Aveiro/Digimedia, Aveiro, Portugal
{lrpc,dora.simoes}@ua.pt
[2] Ferneto S.A., Vagos, Portugal
joao.carapinha@ferneto.com

Abstract. This research work is carried out in a business-to-business context, in a Portuguese multinational and proposes the design, development and validation of an electronic book (ebook) as a business communication tool to promote the consumption of bread for the children's audience. This work justifies the need to fit, based on the literature review on communication tools dedicated to children, the ebook as a new way of promoting a product and a smart learning. We talk about digital communication and communication strategies that appear on the Internet, with a focus on the children's audience. Nowadays, parents spend less time with their children, trying to protect them inside the home, where technological entertainment takes place. Thus, arise marketing strategies that get confused with entertainment, as is the case of eatertainment and advergame. Framed in content marketing, the ebook is a teaching and entertainment tool, by which the children's audience is the most interesting target audience. The methodology used in the present study is development research and the focus group is the method of data collection. The main results obtained are that the cereals best known by the group of children studied are wheat and corn. Children like bread and essentially associate it with sensory experiences (hot, tasty, thick, fluffy, crispy). Children know and eat more types of wheat bread. They usually eat bread twice a day and like to eat it at breakfast. Children do not appreciate the baker profession, so there is an opportunity to promote the baking industry through the presentation of good reasons to be a baker. Regarding the interactive elements most appreciated in an ebook, the results show that children prefer games and quizzes.

Keywords: Digital communication · Children's audience · Eatertainment · Smart learning · Electronic book · Bakery

1 Introduction

This research work is framed in digital marketing, which concerns the promotion of products and brands through the new media. Morais (2011) lists some aspects that distinguish digital marketing from traditional marketing. In digital marketing, a real-time responsiveness to customer needs is required. He or she is the one who decides, being more difficult to influence him or her and strict criteria of segmentation are necessary. In digital environment, it is possible to know the consumer behavior that marketers

Ó. Mealha et al. (eds.), *Citizen, Territory and Technologies: Smart Learning Contexts and Practices*, Smart Innovation, Systems and Technologies 80, DOI 10.1007/978-3-319-61322-2_21

can evaluate by their digital track. As the Internet is the most connected medium of digital marketing, companies today need to carry their campaigns to this new media.

The present work is carried out in a business context, through the proposal of a company which intends to promote the baking sector, using a recent mascot, through a digital content. The company is framed as business-to-business and we are interested in creating something that allows it to approach the final consumer. Thus the children's audience becomes the most interesting target.

Since the book is an educational tool through which it is possible to reach children and considering that everything that is digital fascinates them, the ebook can be a motivating and smart tool for this public. Thus, we chose in this work to design, develop and validate an ebook to meet the needs of the company. Currently, although there are already several ebooks, these still do not stand out in the digital environment (Coutinho and Pestana 2015) and this format has not yet developed all its potentialities, especially in relation to ebooks for children (Baltar 2016), mainly in Portugal.

The research goals are divided between a general one, which relates to the product to be created, as well as several specific ones, taking into account the specificity of the company and its sector. Thus, the design, development and validation of an ebook as a business communication tool, to promote the consumption of bread to the children's audience is the main purpose. In order to achieve this objective, several specific objectives are addressed, namely: (1) figure out what age group from 7 to 9 years old children know about bread and what they associate with it; (2) discover which interactive elements of ebooks facilitate interaction and smart learning; (3) design an ebook that promotes the brand and the baking industry to the target audience through marketing strategies for this target; (4) develop an ebook with interactive elements that make the interaction more fun, while helping to promote the brand; (5) validate the ebook created with the target audience and identify improvements.

2 Digital Communication

The evolution of technology has accompanied marketing, as both have transformed the communication strategies of companies with their customers. According to Ryan (2014), the influence of technology on marketing goes through four phases: (1) the emergence of a new technology; (2) its positioning in the market; (3) analysis of strategies to attract the target audience; and (4) the transformation of technology into conventional. The same author says that, like traditional marketing, the essence of digital marketing is the link between people, with the aim of stimulating sales. Thus, it is more important to understand how consumers use a particular technology than to understand the technology itself and it is crucial to understand its evolution, especially in relation to marketing in the technological age. The so-called marketing 2.0 is directed towards the customer, who becomes more informed and is considered a "consumer 2.0". This type of consumer no longer passively accedes to the media content, starts to control it and has the ability to establish the value of a product according to their own needs. This consumer change is a challenge for marketers and requires the adoption of new smart communication strategies in marketing.

2.1 Strategies

Ryan (2014) argues that the fact that digital media reaches a wider audience than tradi-tional media improves communication with a particular niche. Thus, it is necessary to know the market and the way it uses technology and to plan strategies to use technology to create a long-term connection with the target audience. The author identifies four main digital marketing strategies: (1) social media marketing (2); email marketing; (3) mobile marketing; and (4) content marketing.

Content marketing is a strategy that links marketers and consumers during the period of research, purchase and evaluation (Ryan 2014) and it is important because it addresses the customer's need for information and allows brands to edit content at low cost. There are various types of content, including blog content, ebooks, infographics, videos and photos. Figure 1 shows a matrix of content marketing where various types of content are positioned between different levels of awareness/purchase and emotional/rational, and content can still have the functions of entertaining and educating.

Fig. 1. Matrix of content marketing (adapted from Chaffey and Bosomworth 2013)

Considering the ebook, we can see that it is between the emotional and the rational factor and between the function of entertaining and educating, which justifies approaching the children's audience as the target audience of this specific content.

2.2 Digital Marketing for the Children's Audience

According to Buckingham (2006), the concept of childhood is a social construction and is therefore subjective and changeable over time.

Nowadays, parents spend less time with their children and have greater concern with their well-being and education. Buckingham (2006) points out that this concern moti-vates parents to isolate children from the outside world and protect them within the home, which is provided with technological entertainment. The same author argues that the media are responsible for the disappearance of childhood, because it separate adults and children through the loss of parental authority. Thus, it is believed that the younger ones have greater knowledge of the new technologies than the older ones and this gives them greater authority. In this context, Prensky (2001) distinguishes "digital immigrants" from "digital natives", pointing out that the first one refers to individuals who witnessed the emergence of new technologies and the second one to individuals who were born in the presence of such technologies. However, Prensky (2009) states that in the near future all humanity will be born in the digital age, so the division of digital immigrants/digital natives does not make sense and it is more relevant to speak in digital wisdom. This refers to the wisdom that technologies add to human knowledge and to the human wisdom necessary for the prudent use of technology. Thus, a "digitally improved person" emerges, who can be called digital homo sapiens or digital human (Prensky 2009).

Buckingham (2006) points out that the change of the concept of family, childhood and the emergence of the consumer society have created a new market where children are the protagonists. According to Rozanski (2011) and Valkenburg and Cantor (2001) the child goes through several stages of development as a consumer. In this work, we are interested in the age group from 7 to 9 years old, which is the school stage, when the child begins to read and reaches a stage in which is already aware of the power of the media but is not yet sufficiently awake for marketing and its influence on consumption. At this stage, the child develops preferences for products and brands and also a critical spirit in relation to advertising: she begins to consider details, thinks about the benefits of a product and becomes influential in family consumption decisions.

Taking into account the digital marketing strategies directed to the children's audience, commercial content is easily confused with media content (Fantoni 2014). With the saturation of advertising in traditional media and the new profile of the digital native, it is necessary for marketers to obtain information about this new audience, so that they can interact with them (Cuesta Cambra et al. 2016). This justifies the main strategies of advertisement: advergame and eatertainment.

2.2.1 Advergame and Eatertainment

The concept of advergame is an agglutination of the words advertising and game and is a smart marketing strategy that combines the playfulness of the game with the persuasive intention of marketing. When playing, the player is exposed to advertising content for longer than traditional advertising through rewards, activating cognitive and empirical structures with the brand (Oliveira et al. 2016). In fact, Cuesta Cambra et al. (2016) argue that individuals who enjoy video games tend to remember a brand better than those who use television as their primary means of communication.

The concept of eatertainment is a junction of the word eat with the word entertainment and consists of the use of entertainment in consumption, in the food sector. In this type of marketing strategy, fun is used as a pretext for devaluing nutritional tables. Sometimes this leads to over-consumption of junk food and fast food (Fantoni 2014). Younger children are attracted to this type of food when exposed to media advertising. Elliott (2015) uses the term "promise of fun" to refer to both eatertainment and advergame as strategies adopted by the food industry. The author is in favor of these strategies as playful moments that can improve eating habits. In fact, the promotion of food based on fun, a relatively recent strategy, is more effective than promotion based on nutritional aspects.

In short, the ebook as a tool of content marketing strategy, with the function of entertaining and smart learning, can be an effective means of promoting food for children, because they can know certain foods and acquire good eating habits through entertainment elements.

3   Electronic Book as Digital Communication Tool for Children's Audience

Teixeira and Gonçalves (2015) refer that a book is more than an object, which contains text, images and other elements and is intended for a particular audience. According to Chartier (1998), the act of reading depends on a text, a reader, its know-how and a support in which the text is inserted. The electronic book (or ebook) is a book in digital format, which contains in its name a reference to the printed book, although it may or may not be a digitized version of the same (Teixeira and Gonçalves 2015).

Coutinho and Pestana (2015) present the advantages of the ebook for both the publisher and the reader. The ebook stands out as being more economical than the printed book; as being easier to buy because it does not require the buyer to move to a physical store; by the considerable saving of physical space relative to the traditional book and consequent portability, because in a mobile device it is possible to have several ebooks while it would be more difficult to carry the same number of books in printed format. The ebook also facilitates tasks such as production, distribution and storage, which become more affordable and accessible to more people.

The most popular ebook formats are MOBI (mobile diminutive), PDF (Portable Document Format), ePub (short for electronic publication) and, more recently, Book App (book in application format). Despite the popularity of MOBI, PDF and ePub are currently the most widely used. PDF is a format created by Adobe Systems in the year 1993 and stands out for its adaptation to different devices. Some of its advantages are the possibility of editing documents and creating interactive forms. EPub is an abbreviation for the words electronic publication and was created by the International Digital Publishing Forum. This format works through XHTML, CSS and XML technologies (Coutinho and Pestana 2015). The Book App is a book in application format and is considered a new format for children's picture book. The term came from iOS apps sold on Apple's App Store.

Speaking about mobile devices, the ePub adapts to the screen, while the PDF remains with predefined dimensions, which makes it difficult to read. However, both formats have the potential to add functionality to the text, which improves the reading experience, making it interactive (Coutinho and Pestana 2015). The book app is designed exclusively for each Apple device and does not adapt to different devices, unless the application is redesigned, which requires high production costs (Sargeant 2015). According to Teixeira and Gonçalves (2015) and Teixeira et al. (2015), the Book App is the most effective children's ebook because it has a greater interactive potential than the other formats (non-linear navigation).

Children's books contain essentially image and text, whose reading depends on certain characteristics and cultural background of the reader (Pinto et al. 2013). According to Teixeira et al. (2015), the illustrated book is a type of children's book that, despite presenting text and image, the latter stands out.

The evolution of the book's supports and the appearance of the ebook give way to mobile devices, which stand out for their portability and relative ease of use. These provide sensory experiences through interactive elements. According to Baltar (2016), the ebook is not yet widely explored, since there is a mere adaptation of the text to various formats and different devices, and the potential of illustrated books or interactive books is ignored. In this sense, Nawotka (2016) says that the Book App was a failed initiative due to the high cost of developing this type of book. Alternatively, publishers have opted for web based and cross-platform editions.

Pinto et al. (2013), with the purpose of analyzing several children's ebooks, divide them into five categories: (1) ebook with similar functionalities to the printed book; (2) ebook adapted from a printed book; (3) ebook adapted from an animated film; (4) interactive ebook (linear interaction); (5) interactive ebook (non-linear interaction). Based on the categories of analysis of children's ebooks created by these authors, it was considered the interactive ebook (non-linear interaction) for this research work, because it presents greater interactive potentialities.

4   Methodologic Process

Considering this research work, whose general goal is the design, development and validation of an ebook as a business communication tool to promote the consumption of bread for children, Development Research was considered the most appropriate methodology. Oliveira (2006) defines three moments of development: (1) analysis and evaluation of the situation; (2) conception and realization of the model; and (3) implementation and evaluation. At the first moment, previous studies can be analyzed in order to create a conceptual framework. Thus, in addition to the literature review and documentary analysis of the company, focus groups and brainstorming sessions are conducted. The second moment concerns the design and development of the product, so it is necessary to define the software used in the design, the publishing platform of ebooks chosen for the development and the elements that make up the ebook. Finally, in the third moment, the final version of the ebook is evaluated using new focus group sessions. According to Bryman (2012), the focus group is an interview with several people (at least four elements) on a specific topic and the person leading the focus group is called a moderator or facilitator. A focus group session should be recorded and later transcribed, because it allows the researcher to know who says what and identify potential leaders in the conversation.

The ebook is defined to age group from 7 to 9 years old refers to the age range of primary school, so the age of 7 is the age at which children are expected to be able to read.

The data collection took place in a primary school in Aveiro. Participants were selected through a non-probabilistic sampling method (convenience sampling) by second, third and fourth year teachers so that two male and two female students from each year participated in the study.

In the first part of data collection three focus group sessions were held in a study room, each session with two male and two female students in the second, third and fourth years, in order to find out what the children in the age group from 7 to 9 years old know about bread and what they associate with it. As expected, it took some involvement from the moderator, who needed to encourage students to talk more, in order to get more information. The script was based on some structured questions, with the help of pictures, to facilitate the interaction with students. The questions had the following topics: (1) associations with the word bread; (2) cereals and ingredients used; (3) like or dislike bread; (4) better known and most consumed types of bread; (5) daily consumption; (6) contact with the bakery area and the profession of baker. All sessions were filmed and recorded for later transcription. During the sessions, there were some limitations: student leaders in contrast to students who barely spoke, which was not always possible to control; technical problems, which interrupted one of the sessions and as a result the students became more restless and began to speak at the same time; interruptions of other people who did not know of the occurrence of the session and entered the study room while the session was taking place.

In the second part, a brainstorming session was held in another study room, with only a male and a female student in each year. The goal was to create a story with the contributions of children. For this session, a creative unlock technique called reversal (Michalko 2006) was used to stimulate the children's imagination. For the use of this technique, sheets of paper were distributed with a table formed by two columns, where in the first one the children could write or draw what they thought about a real bread and in the second what they thought about a super bread, a bread with superpowers. They were then asked to create a text on this subject based on their initial ideas about super bread. This session was also filmed and recorded, not only to guarantee the occurrence of the same, but also to record any important interactions for analysis.

The third part was held in a classroom and participated the same number of students who participated in the first part. The children explored the Plataforma Digital dos Concelhos de Portugal (http://www.pdcp.pt/). This is a platform developed by CITI (Centro de Investigação para Tecnologias Interativas), from Universidade Nova de Lisboa. According to the website, it concerns a "collection of original books written and edited for primary school students". The books are composed by text (which can be read and heard), videos, 3D animations, bookmarks, featured vocabulary, quizzes and other interactive elements. A small questionnaire was carried out on the interactive elements that they liked best in one of the ebooks of the platform. As this part required the use of computers and there were only two computers, the main limitation was that only two students could answer the questionnaire at a time, which took longer than expected.

Focus groups and brainstorming were analyzed using content analysis. In the focus group, in order to not identify the students, a code system was created in which the number 2 was assigned to the second year, the number 3 to the third and the number 4 to the fourth, and to each student was assigned a letter from A to D, according to the place where they were sitting. The questionnaire was analyzed based on the occurrence of responses to identify preferred interactive elements.

5   Design and Development

With the data analysis of the three focus group sessions, and taking into account the topics of the questions, we reached the following conclusions:

With regard to the associations with the word bread, the students answered that the bread comes from a bakery; it is something edible; it is tasty and thick; it makes them remember flour, crumbs, seeds, meat, toast, wheat, dough and also Jesus and joy. Speaking about cereals, the students generally know wheat and corn and only in the fourth year they spoke in the rye. In the third and fourth years they recognized the oats through an image. In general, they know what ingredients are used for making bread. However, only in the fourth year they spoke about salt. All the students said they like bread. The main reasons are: to be hot, tasty, fluffy and crispy, to have sticky dough and to have cereals and be healthy. It is important to note that they also talked about butter. About the most well-known types of bread, the students based on wheat bread, talking about whole wheat bread, bijou bread, bread with seeds, bread form, baguette and grandmother's bread. Only in the fourth year they mentioned the rye bread, although in the third year they spoke of dark bread. Although they knew the grandmother's bread, only in the third year they associated the image with the name.

In general, they also eat more wheat bread: grandmother's bread and bread with seeds. Most students eat bread twice a day and the part of the day they eat the most is during the afternoon. However, most students prefer to eat bread at breakfast, because they said that bread tastes better in the morning, because at that time they are hungrier and because the bread is hot. More than half of the students never entered a bakery, in the area where the bread is made. However, about half have made bread in other contexts, either on study visits or at home with family members. The majority showed no interest in the profession of baker, giving several justifications: they had to stand all day; they had to do only one thing; they had to wear an apron and a hat and have to touch the dough, which gets sticky and then they had to wash their hands; they had to wake up very early and work every day and they could become hungry and be expelled for eating the bread.

With the combination of ideas resulting from the brainstorming session, a story was created that we can synthesize as:

> "A baker called Luis created a super bread called Henrique, which was a tiger bread unlike any other. Henry felt bored closed in the Luis bakery and wanted to explore the world. One day he discovered that beside the bakery there was an abandoned factory that had been occupied by baguettes and that the factory had been transformed into their kingdom. The baguettes, which had always been closed in his kingdom, did not know that there were other kinds of bread. Thanks to Henrique, they learned that the world of bread is diverse. Royal Baguette, queen of all baguettes, then gave Henrique a super power: he was going to save breakfast for all the boys and girls in the world."

Regarding the questionnaire, we show Fig. 2, which presents the percentage of answers to the question concerning the interactive elements.

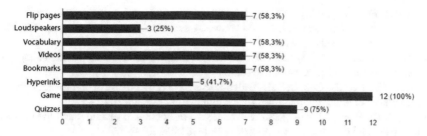

Fig. 2. Interactive elements

As we can see, 100% of students liked the game followed by 75% who liked quizzes and slightly more than half liked the flipping pages, vocabulary, videos and bookmarks.

6 Final Considerations and Future Work

This research work is still in progress. Meanwhile, with all the data obtained, it is possible to make an initial planning of the ebook to be developed, which will begin with a presentation of the mascot of the company, followed by the story about the super bread, which will essentially consist of image and text and will be divided by bookmarks. Each one will relate to each type of bread spoken in history, so the child can simply know the types of bread without following the story linearly. Throughout history there will be some activities that will make it interactive. Considering the tiger bread, which is so called by having "taint" like tiger skin, there could be a zebra bread. Thus, the idea arises of an activity called "What if the bakery was a zoo?", where some animals could name new types of bread. Allied to this idea, we can challenge the child to personalize their bread, through drawings and clippings. Another idea would be an activity called "The bread in the world", where the word bread would appear in several languages, each word accompanied by its sound. Considering that children associate bread with other foods like butter, the idea arises of another activity where food suggestions are presented to accompany with bread and thus try to increase the bread consumption. As the results show that children do not appreciate the baker profession, it is pertinent to present good reasons to be a baker. At the end of the story, there will be a quizz for the child to test their reading interpretation.

In the design phase of the ebook we will use the software Adobe InDesign CC2017, because we consider it the most effective software. In the development phase, we chose the platform Active Textbook, due to the number of adjacent features.

References

Baltar E (2016) La función del editor en la era digital - Desafíos y oportunidades. Telos. Cuad. Comun. e Innovación, pp 1–9

Bryman A (2012) Focus groups. In: Social research methods, 4th edn. Oxford University Press, New York, pp 500–520

Buckingham D (2006) Crescer na era das mídias: após a morte da infância. Florianópolis

Chaffey DD, Bosomworth D (2013) Digital marketing strategy guide. SmartInsights

Chartier R (1998) A aventura do livro: do leitor ao navegador. (UNESP, Ed.): Imprensa Oficial do Estado de São Paulo, São Paulo

Coutinho P, Pestana O (2015) Ebooks: evolução, características e novas problemáticas para o mercado editorial. Páginas a&b 3:169–195

Cuesta Cambra U, Niño González JI, Arroyo Lacunza Ó (2016) Advergaming - La interacción como clave del futuro digital. Telos Cuad Comun e Innovación 104:1–8

Elliott C (2015) Big Food' and "gamified" products: promotion, packaging, and the promise of fun. Crit Public Health 25:2–13. doi:10.1080/09581596.2014.953034

Fantoni A (2014) Estratégias de comunicação publicitária no ambiente online para o público infantil: o caso das marcas Tang e Trakinas. Universidade Federal do Rio Grande do Sul

Michalko M (2006) False Faces. In: Thinkertoys: a handbook of creative-thinking techniques, 2nd edn. Ten Speed Press, pp 43–52

Morais P (2011) Diferenças entre ambiente digital e ambiente tradicional. In: Marketting Digital

Nawotka E (2016) La revolución sigue, pero no triunfa. Telos Cuad Comun e Innovación, 1–4

Oliveira LR (2006) Metodologia do desenvolvimento: um estudo de criação de um ambiente de e- learning para o ensino presencial universitário. Educ Unisinos 10:69–77

Oliveira S, Zagalo N, Melo A (2016) O Advergame como ferramenta publicitária: um estudo exploratório. Comun Cult e Estratégias, 152–175

Pinto AL, Zagalo N, Coquet E (2013) Pedra, papel ou digital: onde lê, como lê e o que lê a criança na era digital. Atas do 9º Encontro Nac (7º Int Investig em Leitura, Lit Infant e Ilus, pp 217–240

Prensky M (2001) Digital natives, digital immigrants. Horiz 9:1–15

Prensky M (2009) H. Sapiens Digital: from digital immigrants and digital natives to digital wisdom. J Online Educ 5:1–11. www.innovateonline.info/index.php?view=article&id=705

Rozanski J (2011) Publicidade infantil: análise de estratégias de comunicação em comerciais de alimentos. Universidade Federal do Rio Grande do Sul

Ryan D (2014) Understanding digital marketing: marketing strategies for engaging the digital generation, 3rd edn. Kogan Page, London

Sargeant B (2015) What is an ebook? What is a book app? And why should we care? An analysis of contemporary digital picture books. Child Lit Educ 46:454–466. doi:10.1007/s10583-015-9243-5

Teixeira DJ, Gonçalves SB (2015) Ebook interativo de histórias infantis: a potencialidade expressiva das narrativas digitais. 1º Congr Nac Novas Narrat encontro Narrat Comun e artes, pp 1–14

Teixeira DJ, Vargas Nunes J, Gonçalves SB, Perassi Luiz de Sousa R (2015) Linguagem visual e princípios de design em ebook interativo infantil. Palíndromo 6:129–143. doi: 10.5965/2175234606122014129

Valkenburg PM, Cantor J (2001) The development of a child into a consumer. J Appl Dev Psychol 22:61–72. doi:10.1016/S0193-3973(00)00066-6

A Review of Proxemics in 'Smart Game-Playing'

Liliana Vale Costa[✉], Ana Isabel Veloso, and Óscar Mealha

Department of Communication and Art, CIC Digital/DigiMedia,
University of Aveiro, Aveiro, Portugal
{lilianavale,aiv,oem}@ua.pt

Abstract. Recent developments in the game industry and in the paradigm of Internet of Things (IoT) have heightened the need for developing innovative solutions to foster movement-based interactions and bring people together in both physical and digital (phygital) environments. Although the existing knowledge on interaction design in game experience is quite extensive, little is known about proxemics in game design and how it can be explored to conceive "smart game-playing". This paper reports on the use of proxemics in digital games and its utility in enhancing game-mediated interactions applied to 'smart ecosystems'. Eight papers published between 2003 and 2016 in English-language publications related with proxemics in digital games were reviewed and met inclusion criteria. This review presents a set of recommendations for applying proxemics in game-mediated interactions and; discusses its role in enabling informational literacy to foster smart learning ecosystems.

Keywords: Proxemics · Learning · Game-mediated interactions · Design recommendations · Smart learning ecosystems · Phygital place

1 Introduction

The advent of the paradigm of Internet of Things (IoT) and current policies regarding smart cities [e.g. (Albino et al. 2015; Caragliu et al. 2011)] have led to a renewed interest in studying citizens' spatial behaviours, their use of shared spaces, artefacts and such strategies to bring people together in public spaces with a gamification approach and location-based games, ingredients that constitute a phygital place (Giovannella et al. n.d.). In addition, changes in the game industry have also been occurring in order to meet societal challenges and needs of new target groups – *i.e.* player-citizens from all ages and of different genders and backgrounds (*e.g.* 'games for a change', 'games with a purpose', 'gamification', 'alternate reality games', 'location-based games', 'mixed-reality games', 'mobile games').

Although there is a growing body of literature that recognises the importance of space in digital games [*e.g.* (Nitsche 2008; Pearce 2008)], there has been little discussion [e.g. (Greenberg 2011; Mueller et al. 2014)] about proxemics, Hall's (1969) theory on a "Hidden Dimension" or a "Silent Language" that humans use to manage their relational distance, applied in game-mediated interactions. Furthermore, the implications of these interactions in addressing the main learning needs and urgency of acquiring information competences in a smart ecosystem have been understudied.

© Springer International Publishing AG 2018
Ó. Mealha et al. (eds.), *Citizen, Territory and Technologies: Smart Learning Contexts and Practices*, Smart Innovation, Systems and Technologies 80, DOI 10.1007/978-3-319-61322-2_22

The goal of this study is to understand the potential of Hall's (1969) proxemics in game-mediated interactions to foster citizen-player driven learning and decision-making toward the community and cities, the basis of a smart game-playing ecosystem. Specifically, we identify a set of recommendations for applying proxemics in phygital game-mediated interactions and discuss its role in enabling informational literacy and learning that is often the basis of individuals' informed choices in daily life.

2 Proxemics in 'Smart Game-Playing'

In the light of such challenges as growing population, resource management, unemployment, mobility and infocommunication overload, it is becoming extremely important to provide citizens with information competences and motivate them to contribute to an ecological, social and economic sustainable environment (Holling 2001).

According to the Timisoara declaration, 'smart environments' are those in which (Giovannella et al. n.d, p. 3): "individuals that take part in the local processes have a high level of skills and, at the same time, are also strongly motivated and engaged by continuous and adequate challenges, provided that their primary needs are reasonably satisfied." In this definition, three game elements can be highlighted and used to introduce the concept of 'smart game-playing' - level of progression and collection of skills; rewarding motivated actions; and involvement of citizens in challenges/missions that can have an impact on society and improve citizens' wellbeing and quality of life.

In fact, game-mediated interactions can foster citizen-player driven learning by bringing people together through shared spaces and artefacts, encouraging the discovery of patterns, engaging citizens with storytelling, and motivating decision-making and changes in behaviors, pro-game instances that occur in an informal social sharing/learning context.

According to Hall (1969), there are four types of space: (a) the intimate space; (b) personal space; (c) social space; and (d) public space. Relative to a specific game instance in a pre-determined context, these "proxemic spaces" can be generated with the position of players and devices, size of devices, number of players, type of game and feedback. In fact, game designers should take these factors into account, when designing different types of mediated interactions, the basic structure and relation of "place" and "digital artefacts and algorithms", constituents of the phygital place.

3 Article Coding, Review and Analysis

We searched peer-reviewed papers in Web of Science, Scopus and Springer. The inclusion criteria used to select candidate papers were: (a) published between 2003 and 2016 and; (b) discussing proxemics in game-mediated interactions. Exclusion criteria were any of the following: (a) duplicate publication; (b) non-English or Portuguese language publications and; (c) not meeting any of the inclusion criteria. Studies that covered proxemics in virtual environments that were not a game (i.e. simulators, *second life...*) were also excluded from the analysis. We used the search terms: (proxemics AND (game OR digital AND game OR digital AND gam* OR videogame OR video AND game OR play OR gamification)). This search yielded 673 potentially eligible papers. From these papers, duplicate papers were removed, with no full access, non-English or non-Portuguese and those neither related with proxemics nor digital games, which yielded a sample of 8. We coded the papers with the following codes: (a) type of games; and (b) recommendations for applying proxemics in game-mediated interactions.

The papers were read and the design features identified in order to compile sets of recommendations for designing proxemics in game-mediated interactions (Table 1). Each study was repeatedly read and recommendations were highlighted using open coding (Given 2008).

Table 1. Type of games and recommendations for applying proxemics in game-mediated interactions

Game type	Recommendations for applying proxemics
Music game with a tangible user interface (Grønbæk et al. 2016)	- Provide immediate feedback in order to enhance the sense of impact of an action - Enable players to move freely and encourage them to share an intimate or personal space through f-formations around tangible devices
Virtual reality game (with head-mounted display) (Economic game) (McCall and Singer 2015)	- Design proxemics in game-mediated interactions accordingly with the level of fairness
A collaborative mediated body-space game with Sony Move Controller (Intangle) (Garner et al. 2014)	- Encourage collaboration between players through the use of shared controllers - Create awareness of time and game space through immediate feedback – i.e. visual, auditory, haptic light, vibration - Foster collaboration, empathy and inclusivity through personally mediated body-space in games
Movement-based game with Move Controller (i-dentity) (Garner et al. 2013)	- Take into account movement when designing game-mediated interactions -Represent the game movements through the use of on-screen avatars, light and sound -Direct the attention of players to other players and viewers by using the physical space - Design the space between players, between players and screens, their movements and social interactions - Enable passive game-playing ('spectator involvement') and encourage interpersonal synchrony (movement-coordinated tasks)

(continued)

Table 1. (*continued*)

Game type	Recommendations for applying proxemics
Augmented reality in public space (Grubert et al. 2012)	- Encourage people's interactions in public space and transportation through the use of augmented reality games
3D cave game (eXperience Induction machine) (Inderbitzin et al. 2009)	- Encourage players' full body interactions in mixed reality - Design game proxemics that vary in winning and losing strategies and game conflict (cooperation vs competition)
Tangible user interface game (Musical embrace) (Mueller et al. 2014)	- Enable passive game-playing (stimulate 'spectator involvement') - Extend the traditional proxemics proposed by Hall (1969) with 'wireless zones'
Many types of games (Rehm et al. 2005)	- Take into account the following aspects: gaze and eye contact, voice volume, posture, body language, cultural taboos about touch and contextual factors and players' implicit actions - Show the initial state and information relative to the positions and movements of the player as well as enhance security and privacy - Transfer different information "from up close" and "from afar" and enable 'passive' game playing

The majority of studies (Table 1) discuss the use of proxemics in game-mediated interactions that occur in the same physical space [e.g. (Garner et al. 2013; Garner et al. 2014; Grønbæk et al. 2016; Inderbitzin et al. 2009; Mueller et al. 2014)]. The scenarios presented in the studies reviewed (Table 1) seem to converge in a number of aspects that can have impact on proxemics, i.e. there is a shared ownership of the physical space (Garner et al. 2013; Garner et al. 2014, Grønbæk et al. 2016; Inderbitzin et al. 2009; Mueller et al. 2014), objects (Grønbæk et al. 2016) and controllers (Garner et al. 2014); an individual ownership of the digital space and both physical and digital representations of the self (Garner et al. 2013; Inderbitzin et al. 2009; McCall and Singer 2015; Mueller et al. 2014, Rehm et al. 2005); turn-based and synchronous interactions (Garner et al. 2013; Garner et al. 2014; Grønbæk et al. 2016; McCall and Singer 2015); the focus is divided into off-screen and on-screen spaces (Garner et al. 2013; Garner et al. 2014; Grønbæk et al. 2016; Grubert et al. 2012; Inderbitzin et al. 2009; Rehm et al. 2005); and the output controllers (e.g. command shaking) are used to transmit feedback (Garner et al. 2013; Garner et al. 2014; Grubert et al. 2012; Rehm et al. 2005).

Thus, we divided 'hot mediated interactions' (the ones established in the same physical space - phygital place) and 'cool mediated interactions' (the ones established in different physical spaces - non-phygital place) in order to propose a model of proxemics in game-mediated interactions (Fig. 1).

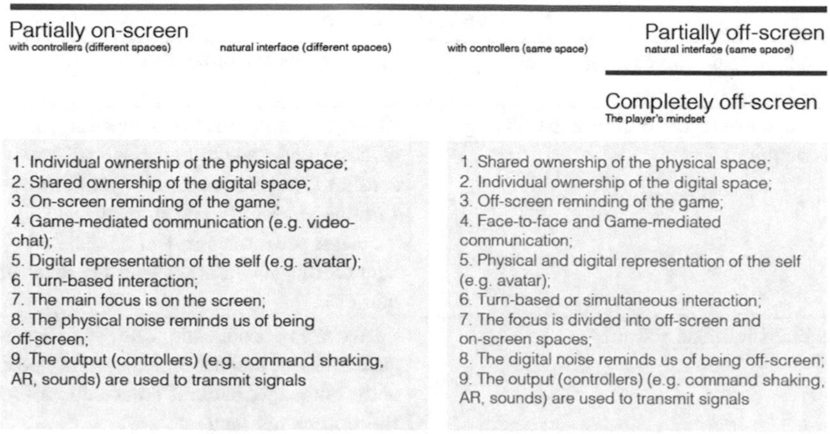

Partially on-screen
with controllers (different spaces)   natural interface (different spaces)   with controllers (same space)   natural interface (same space)

Partially off-screen

Completely off-screen
The player's mindset

1. Individual ownership of the physical space;	1. Shared ownership of the physical space;
2. Shared ownership of the digital space;	2. Individual ownership of the digital space;
3. On-screen reminding of the game;	3. Off-screen reminding of the game;
4. Game-mediated communication (e.g. video-chat);	4. Face-to-face and Game-mediated communication;
5. Digital representation of the self (e.g. avatar);	5. Physical and digital representation of the self (e.g. avatar);
6. Turn-based interaction;	6. Turn-based or simultaneous interaction;
7. The main focus is on the screen;	7. The focus is divided into off-screen and on-screen spaces;
8. The physical noise reminds us of being off-screen;	8. The digital noise reminds us of being off-screen;
9. The output (controllers) (e.g. command shaking, AR, sounds) are used to transmit signals	9. The output (controllers) (e.g. command shaking, AR, sounds) are used to transmit signals

Fig. 1. Model of Proxemics in game-mediated interactions

As can be observed in Fig. 1, the distance between different players increases (blue and green colour) as the interaction is more mediated. By contrast, players get closer when they share the same space and their interaction is as natural as possible (yellow and red colour).

4 The Potential of Proxemics in 'Smart Game-Playing'

Games are a source of information and communication that can have an impact on daily lives, routines, homes and cities. As games can function as literate environments in a smart ecosystem, we present the requirements defined by UNESCO (2011) to create a literate environment and the main implications for proxemics (Table 2).

Table 2. Implications of 'smart game-playing' for proxemics

Requirements for developing literate environments	Implications for proxemics
- Transfer the literacy skills to daily life activities and tasks	- Encourage people's interactions in public space through the use of augmented reality games
- Invite the participation of individuals and the use of their literacy skills to interact with the environment	- Encourage players' full body interactions in mixed-reality - Enable passive game-playing (stimulate 'spectator involvement') - Provide immediate feedback in order to enhance the sense of impact of an action

(continued)

Table 2. (*continued*)

Requirements for developing literate environments	Implications for proxemics
- Give a sense of control over the literacy skills acquired	- Design game proxemics that vary in winning and losing strategies and game conflict (cooperation vs competition) - Enhance security and privacy - Design proxemics in game-mediated interactions accordingly with the level of fairness
- Required intrinsic rewards	- Take into account the following aspects: gaze and eye contact, voice volume, posture, body language, cultural taboos about touch and contextual factors;
- Provide opportunities to participate in community	- Encourage people's interactions in public space through the use of augmented reality games - Enable players to move freely and encourage them to share an intimate/personal space through f-formations around tangible devices - Encourage collaboration between players through the use of shared controllers - Foster collaboration, empathy and inclusivity through personally mediated body-space in games

5 Discussion

In this paper, a set of recommendations were proposed to encourage game-mediated interactions with an explicit and intentional logical use of space and we discuss in what way these can meet some of the requirements established for developing 'literate environments' according to UNESCO. A 'smart game-playing' consists in using such game strategies as level of progression and collection of skills; rewarding motivated actions; and involvement of citizens in challenges/missions in order to have an impact on society and improve citizens' wellbeing and quality of life, shown in the proposed model of proxemics in game play. For example, in a shared urban space, proxemics in game-mediated interactions can create tension between idealized mental representations and the environment. Indeed, depending on citizen-players' interactions with the environment (i.e. whether the human body is the reference point or a camera in order to frame and narrate a perspective) (Nitsche 2008), their position and perception toward the environment also tend to change. In this reciprocal affordance (Gibson 1986) between what the player can afford from the environment and vice-versa, the design of proxemics in game-mediated interaction beyond game conventions (perceived affordances) (Norman 2013) is crucial to invite citizen-players to (re) act to an adaptive and constantly-changing ecosystem with the progress of technologies. This review, however, has some limitations, so the results need to be interpreted with caution. Only peer-reviewed papers were analysed and recommendations proposed by enterprises or other entities could help to understand the potential of proxemics in these interactions. Future research should analyse the proposed model for different types of devices and different goals to be achieved in a 'smart ecosystem.'

Acknowledgments. Research reported in this publication was supported by FCT (Fundação para a Ciência e a Tecnologia) and ESF (European Social Fund) under Community Support Framework III SFRH/BD/101042/2014 and also supported by the grant SFRH/BSAB/ 128152/2016 (Fundo Social Europeu and Portuguese financial resources from the Ministry of Science, Technology and Higher Education - MCTES).

References

Albino V, Berardi U, Dangelico RM (2015) Smart cities: definitions, dimensions, performance, and initiatives. J Urban Technol 22:3–21. doi:10.1080/10630732.2014.942092

Aslerd (2016) Timisoara declaration on better learning for a better world through people centred smart learning ecosystems, pp 1–9

Caragliu A, Del Bo C, Nijkamp P (2011) Smart cities in Europe. J Urban Technol 18:65–82. doi:10.1080/10630732.2011.601117

Garner J, Wood G, Danilovic S, et al (2014) Intangle: exploring interpersonal bodily interactions through sharing controllers. In: Proceedings of the first ACM SIGCHI annual symposium on computer-human interaction in play, pp 413–414

Garner J, Wood G, Pijnappel S, et al (2013) Combining moving bodies with digital elements. In: Proceedings of the 9th Australasian conference on interactive entertainment matters of life and death - IE 2013, pp 1–10. doi:10.1145/2513002.2513014

Gibson JJ (1986) The ecological approach to visual perception. J Soc Architect Historians 39:332. doi:10.2307/989638

Giovannella C, Jansen D, Maillet K, Texeira A, Vasiu R, Koch G Timosara declaration Better Learning for a Better World through People Centred Smart Learning Ecosystems, 19th of May 1–9

Given LM (2008) The sage encyclopedia of qualitative research methods, p 1043. MIT Press. doi:10.4135/9781412963909

Greenberg S (2011) Opportunities for proxemic interactions in ubicomp (keynote). In: Campos P, Graham N, Jorge J, Nunes N, Palanque P, Winckler M (eds) Human-Computer Interaction – INTERACT 2011. INTERACT 2011. Lecture Notes in Computer Science, vol 6946, pp 3–10. Springer, Heidelberg

Grønbæk JE, Jakobsen KB, Petersen MG, Rasmussen MK, Winge J, Stougaard J (2016) Designing for children's collective music making: how spatial orientation and configuration matter. In: Proceedings of the 9th nordic conference on human-computer interaction. ACM, New York. doi:10.1145/2971485.2971552

Grubert J, Morrison A, Munz H, Reitmayr G (2012) Playing it real : magic lens and static peephole interfaces for games in a public space. In: Proceedings of MobileHCI 2012, vol 10. doi:10.1145/2371574.2371609

Hall E (1969) The hidden dimension. Anchor Books, New York

Holling CS (2001) Understanding the complexity of economic, ecological, and social systems. Ecosystems 4:390–405. doi:10.1007/s10021-001-0101-5

Inderbitzin M, Wierenga S, Väljamäe A, Bernardet U, Verschure PF (2009) Social cooperation and competition in the mixed reality space eXperience induction machine XIM. In: virtual reality, pp 153–158

McCall C, Singer T (2015) Facing off with unfair others: introducing proxemic imaging as an implicit measure of approach and avoidance during social interaction. PLoS ONE. doi:10. 1371/journal.pone.0117532

Mueller F, Stellmach S, Greenberg S, et al (2014) Proxemics play: understanding proxemics for designing digital play experiences. In: Proceedings of the 2014 conference on designing interactive systems - DIS 2014, pp 533–542. doi:10.1145/2598510.2598532

Nitsche M (2008) Video game spaces: image, play, and structure in 3D worlds. MIT Press, London

Norman DA (2013) The design of everyday things: revised and expanded edition. Basic Books, New York

Pearce C (2008) Spatial literacy: reading (and writing) game space. In: Proceedings future and reality of gaming (FROG), 17–19 October

Rehm M, André E, Nischt M (2005) Let's come together—social navigation behaviors of virtual and real humans. In: Maybury M, Stock O, Wahlster W (eds) International conference on intelligent technologies for interactive entertainment, pp 124–133. Springer, Heidelberg. doi:10.1007/11590323_13

UNESCO (2011) Creating and sustaining literate environments. UNESCO Bangkok Asia and Pacific Regional Bureau for Education, Bangkok, Thailand. ISBN 978-92-9223-379-2

Author Index

© Springer International Publishing AG 2018   227
Ó. Mealha et al. (eds.), *Citizen, Territory and Technologies: Smart Learning Contexts and Practices*,
Smart Innovation, Systems and Technologies 80, DOI 10.1007/978-3-319-61322-2

Printed in the United States
By Bookmasters